PE EXAM PREP

MW00814786

ELECTRICAL ENGINEERING
PE LICENSE REVIEW

Ninth Edition

Lincoln D. Jones, MS, PE Elec. Eng.

KAPLAN) AEC EDUCATION

This publication is designed to provide accurate and authoritative information in regard to the subject matter covered. It is sold with the understanding that the publisher is not engaged in rendering legal, accounting, or other professional service. If legal advice or other expert assistance is required, the services of a competent professional person should be sought.

President: Roy Lipner
Vice President & General Manager: David Dufresne
Vice President of Product Development and Publishing: Evan M. Butterfield
Editorial Project Manager: Laurie McGuire
Director of Production: Daniel Frey
Production Editor: Caitlin Ostrow
Creative Director: Lucy Jenkins

Published by Kaplan AEC Education
30 South Wacker Drive
Chicago, IL 60606-7481
(312) 836-4400
www.kaplanaecengineering.com

CONTENTS

Introduction

HOW TO USE THIS BOOK

Electrical Engineering PE License Review and its companion texts form a three-step approach to preparing for the Principles and Practice of Engineering (PE) exam:

■ *Electrical Engineering PE License Review* contains the conceptual review of electrical engineering topics for the exam, including key terms, equations, analytical methods and reference data. Because it does not contain problems and solutions, the book can be brought into the open-book PE exam as one of your references.

■ *Electrical Engineering PE Problems & Solutions* provides problems for you to solve in order to test your understanding of concepts and techniques. Ideally, you should solve these problems after completing your conceptual review. Then, compare your solution to the detailed solutions provided, to get a sense of how well you have mastered the content and what topics you may want to review further.

■ *Electrical & Computer Engineering PE Sample Exam* provides complete morning and afternoon exam sections so that you can simulate the experience of taking the PE test within its actual time constraints and with questions that match the test format. Take the sample exam after you're satisfied with your review of concepts and problem-solving techniques, to test your readiness for the real exam.

BECOMING A PROFESSIONAL ENGINEER

To achieve registration as a professional engineer there are four distinct steps: (1) education, (2) the Fundamentals of Engineering/Engineer-In-Training (FE/EIT) exam, (3) professional experience, and (4) the professional engineer (PE) exam, more formally known as the Principles and Practice of Engineering Exam. These steps are described in the following sections.

Education

The obvious appropriate education is a B.S. degree in electrical engineering from an accredited college or university. This is not an absolute requirement. Alternative, but less acceptable, education is a B.S. degree in something other than electrical engineering, or a degree from a non-accredited institution, or four years of education but no degree.

Fundamentals of Engineering (FE/EIT) Exam

Most people are required to take and pass this eight-hour multiple-choice examination. Different states call it by different names (Fundamentals of Engineering, E.I.T., or Intern Engineer), but the exam is the same in all states. It is prepared and graded by the National Council of Examiners for Engineering and Surveying (NCEES). Review materials for this exam are found in other Engineering Press books such as *Fundamentals of Engineering: FE/EIT Exam Preparation.*

Experience

Typically one must have four years of acceptable experience before being permitted to take the Professional Engineer exam. Both the length and character of the experience will be examined. It may, of course, take more than four years to acquire four years of acceptable experience.

Professional Engineer Exam

The second national exam is called Principles and Practice of Engineering by NCEES, but just about everyone else calls it the Professional Engineer or PE exam. All states, plus Guam, the District of Columbia, and Puerto Rico, use the same NCEES exam.

ELECTRICAL ENGINEERING PROFESSIONAL ENGINEER EXAM

The reason for passing laws regulating the practice of civil engineering is to protect the public from incompetent practitioners. Most states require engineers working on projects involving public safety to be registered, or to work under the supervision of a registered professional engineer. In addition, many private companies encourage or require engineers in their employ to pursue registration as a matter of professional development. Engineers in private practice, who wish to consult or serve as expert witnesses, typically also must be registered. There is no national registration law; registration is based on individual state laws and is administered by boards of registration in each of the states. You can find a list of contact information for and links to the various state boards of registration at the Kaplan AEC Web site: *www.kaplanaecengineering.com*. This list also shows the exam registration deadline for each state.

Examination Development

Initially the states wrote their own examinations, but beginning in 1966 the NCEES took over the task for some of the states. Now the NCEES exams are used by all states. This greatly eases the ability of an engineer to move from one state to another and achieve registration in the new state.

The development of the engineering exams is the responsibility of the NCEES Committee on Examinations for Professional Engineers. The committee is composed of people from industry, consulting, and education, plus consultants and subject matter experts. The starting point for the exam is a task analysis survey, which NCEES does at roughly 5- to 10-year intervals. People in industry, consulting, and education are surveyed to determine what electrical engineers do and what knowledge is needed. From this NCEES develops what it calls a "matrix of knowledge" that forms the basis for the exam structure described in the next section.

The actual exam questions are prepared by the NCEES committee members, subject matter experts, and other volunteers. All people participating must hold professional registration. Using workshop meetings and correspondence by mail, the questions are written and circulated for review. Although based on an understanding of engineering fundamentals, the problems require the application of practical professional judgment and insight.

Examination Structure

The morning breadth exam consists of 40 multiple-choice questions covering the following areas of electrical engineering (relative exam weight for each topic is shown in parentheses):

■ Basic electrical engineering (45%). This area encompasses economics, ethics, professional practice, safety, electric circuits, electric and magnetic theory and applications, and digital logic.

■ Electronics, electronic circuits and components (20%).

■ Controls and communication systems (15%)

■ Power (20%)

You will have four hours to complete the breadth exam.

The afternoon depth exam is actually three exams; you choose the depth exam you wish to take. The depth exams are:

■ Computers

■ Electronics, Controls, and Communications

■ Power

Clearly, you should choose the exam that best matches your training and professional practice. You will have four hours to answer the 40 multiple-choice questions that make up the depth exam.

Both the breadth and depth questions include four possible answers (A, B, C, D) and are objectively scored by computer.

For more information on the topics and subtopics and their relative weights on the breadth and depth portions, visit the NCEES Web site at *www.ncees.org*.

Exam Dates

The National Council of Examiners for Engineering and Surveying (NCEES) prepares Professional Engineer exams for use on a Friday in April and October of each year. Some state boards administer the exam twice a year in their state, whereas others offer the exam once a year. The scheduled exam dates for the next ten years can be found on the NCEES Web site (*www.ncees.org/exams/schedules/*).

People seeking to take a particular exam must apply to their state board several months in advance.

Exam Procedure

Before the morning four-hour session begins, proctors will pass out an exam booklet, answer sheet, and mechanical pencil to each examinee. The provided pencil is the only writing instrument you are permitted to use during the exam. If you need an additional pencil during the exam, a proctor will supply one.

Fill in the answer bubbles neatly and completely. Questions with two or more bubbles filled in will be marked as incorrect, so if you decide to change an answer, be sure to erase your original answer completely.

The afternoon session will begin following a one-hour lunch break.

In both the morning and afternoon sessions, if you finish more than 15 minutes early you may turn in your booklet and answer sheet and leave. In the last 15 minutes, however, you must remain to the end of the exam in order to ensure a quiet environment for those still working and an orderly collection of materials.

Preparing for and Taking the Exam

Give yourself time to prepare for the exam in a calm and unhurried way. Many candidates like to begin several months before the actual exam. Target a number of hours per day or week that you will study, and reserve blocks of time for doing so. Creating a review schedule on a topic-by-topic basis is a good idea. Remember to allow time for both reviewing concepts and solving practice problems.

In addition to review work that you do on your own, you may want to join a study group or take a review course. A group study environment might help you stay committed to a study plan and schedule. Group members can create additional practice problems for one another and share tips and tricks.

You may want to prioritize the time you spend reviewing specific topics according to their relative weight on the exam, as identified by NCEES, or by your areas of relative strength and weakness.

People familiar with the psychology of exam taking have several suggestions for people as they prepare to take an exam.

1. Exam taking involves really, two skills. One is the skill of illustrating knowledge that you know. The other is the skill of exam taking. The first may be enhanced by a systematic review of the technical material. Exam-taking skills, on the other hand, may be improved by practice with similar problems presented in the exam format.

2. Since there is no deduction for guessing on the multiple choice problems, answers should be given for all of them. Even when one is going to guess, a logical approach is to attempt to first eliminate one or two of the four alternatives. If this can be done, the chance of selecting a correct answer obviously improves from 1 in 4 to 1 in 3 or 1 in 2.

3. Plan ahead with a strategy. Which is your strongest area? Can you expect to see several problems in this area? What about your second strongest area? What is your weakest area?

4. Plan ahead with a time allocation. Compute how much time you will allow for each of the subject areas in the breadth exam and the relevant topics in the depth exam. You might allocate a little less time per problem for those areas in which you are most proficient, leaving a little more time in subjects that are difficult for you. Your time plan should include a reserve block for especially difficult problems, for checking your scoring sheet, and to make last-minute guesses on problems you did not work. Your strategy might also include time allotments for two passes through the exam—the first to work all problems for which answers are obvious to you, and the second to return to the more complex, time-consuming problems and the ones at which you might need to guess. A time plan gives you the confidence of being in control and keeps you from making the serious mistake of misallocation of time in the exam.

5. Read all four multiple-choice answers before making a selection. An answer in a multiple-choice question is sometimes a plausible decoy—not the best answer.

6. Do not change an answer unless you are absolutely certain you have made a mistake. Your first reaction is likely to be correct.

7. Do not sit next to a friend, a window, or other potential distractions.

Exam Day Preparations

The exam day will be a stressful and tiring one. This will be no day to have unpleasant surprises. For this reason we suggest that an advance visit be made to the examination site. Try to determine such items as

1. How much time should I allow for travel to the exam on that day? Plan to arrive about 15 minutes early. That way you will have ample time, but not too much time. Arriving too early, and mingling with others who also are anxious, will increase your anxiety and nervousness.

2. Where will I park?

3. How does the exam site look? Will I have ample workspace? Where will I stack my reference materials? Will it be overly bright (sunglasses), cold (sweater), or noisy (earplugs)? Would a cushion make the chair more comfortable?

4. Where are the drinking fountains and lavatory facilities?

5. What about food? Should I take something along for energy in the exam? A bag lunch during the break probably makes sense.

What to Take to the Exam

The NCEES guidelines say you may bring only the following reference materials and aids into the examination room for your personal use:

1. Handbooks and textbooks, including the applicable design standards.

2. Bound reference materials, provided the materials remain bound during the entire examination. The NCEES defines "bound" as books or materials fastened securely in their covers by fasteners that penetrate all papers. Examples are ring binders, spiral binders and notebooks, plastic snap binders, brads, screw posts, and so on.

3. A battery-operated, silent, nonprinting, noncommunicating calculator from the NCEES list of approved calculators. For the most current list, see the NCEES Web site (*www.ncees.org*). You also need to determine whether or not your state permits preprogrammed calculators. Bring extra batteries for your calculator just in case; many people feel that bringing a second calculator is also a very good idea.

At one time NCEES had a rule barring "review publications directed principally toward sample questions and their solutions" in the exam room. This set the stage for restricting some kinds of publications from the exam. *State boards may adopt the NCEES guidelines, or adopt either more or less restrictive rules.* Thus an important step in preparing for the exam is to know what will—and will not—be permitted. We suggest that if possible you obtain a written copy of your state's policy for the specific exam you will be taking. Occasionally there has been confusion at individual examination sites, so a copy of the exact applicable policy will not only allow you to carefully and correctly prepare your materials, but will also ensure that the exam proctors will allow all proper materials that you bring to the exam.

As a general rule we recommend that you plan well in advance what books and materials you want to take to the exam. Then they should be obtained promptly so you use the same materials in your review that you will have in the exam.

License Review Books

The review books you use to prepare for the exam are good choices to bring to the exam itself. After weeks or months of studying, you will be very familiar with their organization and content, so you'll be able to quickly locate the material you want to reference during the exam. Keep in mind the caveat just discussed—some state boards will not permit you to bring in review books that consist largely of sample questions and answers.

Textbooks

If you still have your university textbooks, they are the ones you should use in the exam, unless they are too out of date. To a great extent the books will be like old friends with familiar notation.

Bound Reference Materials

The NCEES guidelines suggest that you can take any reference materials you wish, so long as you prepare them properly. You could, for example, prepare several volumes of bound reference materials, with each volume intended to cover a particular category of problem. Maybe the most efficient way to use this book would be to cut it up and insert portions of it in your individually prepared bound materials. Use tabs so that specific material can be located quickly. If you do a careful and systematic review of civil engineering, and prepare a lot of well-organized materials, you just may find that you are so well prepared that you will not have left anything of value at home.

Other Items

In addition to the reference materials just mentioned, you should consider bringing the following to the exam:

- *Clock*—You must have a time plan and a clock or wristwatch.

- *Exam assignment paperwork*—Take along the letter assigning you to the exam at the specified location. To prove you are the correct person, also bring something with your name and picture.

- *Items suggested by advance visit*—If you visit the exam site, you probably will discover an item or two that you need to add to your list.

- *Clothes*—Plan to wear comfortable clothes. You probably will do better if you are slightly cool.

- *Box for everything*—You need to be able to carry all your materials to the exam and have them conveniently organized at your side. Probably a cardboard box is the answer.

Examination Scoring and Results

The questions are machine-scored by scanning. The answers sheets are checked for errors by computer. Marking two answers to a question, for example, will be detected and no credit will be given.

Your state board will notify you whether you have passed or failed roughly three months after the exam. Candidates who do not pass the exam the first time may take it again. If you do not pass you will receive a report listing the percentages of questions you answered correctly for each topic area. This information can help focus the review efforts of candidates who need to retake the exam.

The PE exam is challenging, but analysis of previous pass rates shows that the majority of candidates do pass it the first time. By reviewing appropriate concepts and practicing with exam-style problems, you can be in that majority. Good luck!

Fundamental Concepts and Techniques

This chapter is presented to help jog your memory regarding some of the material found in the electrical engineering portion of the Fundamentals of Engineering (FE) examination and relevant to the Principles and Practice of Engineering (PE) exam. Also, this chapter includes some new terminology that was not in use several years ago (for instance, the voltage drop across an ammeter is now referred to as "burden voltage").

The two main topics reviewed in this chapter are circuits (and associated topics) and electrostatic and magnetic fields. You will need a good understanding of the following types of problems, which are likely to appear on the exam:

■ Transient analysis

■ Impedance matching

■ Meters and waveforms

■ Resonance and bandwidth

This chapter will review these topics as well.

You probably already have a solid understanding of conventional dc and ac circuit analysis problems involving the use of Kirchhoff's voltage and current laws, along with Thevenin's and Norton's theorems. If you feel you need to refresh your knowledge of these topics, the references at the end of this chapter are good resources. This book will not review those topics directly.

BASIC CIRCUIT ANALYSIS

Several practice problems on circuits follow. First, however, we will look briefly at a circuit with a requirement for maximum power transfer, and then a conversion to a dependent current source.

Suppose you need to determine the maximum power that can be dissipated in a variable load resistor nested in a multiple source circuit. First, you must know the size of the resistor. The **maximum power transfer theorem** (for dc circuits) states that: maximum power is extracted from a circuit when the circuit is converted to Thevenin's equivalent and the load resistance is just equal to Thevenin's resistance. Another way of stating this maximum power transfer theorem is to say that half of the power is dissipated in the load and the other half in Thevenin's equivalent resistance. Consider the circuit of Figure 1.1; all that is necessary is to isolate the load resistor from being "buried" in the circuit. This is easily done by rearranging the circuit. One way is to move the load resistor to the right side of the circuit, and then find Thevenin's equivalent circuit for that portion on the left.

For the rearranged circuit of Figure 1.1b, the open circuit voltage (*i.e.*, with R_L temporarily disconnected) will be found by the node method of analysis ($V_{OC} = V_{Thev}$):

$$\sum I\text{'s at } V_2: \quad \frac{(V_2 - V_1)}{2} + \frac{(V_2 - V_3)}{4} = 0$$

$$\sum I\text{'s at } V_3: \quad \frac{(V_3 - V_2)}{4} - 2 + \frac{V_3}{6} = 0$$

Solving for $V_2 = V_{OC}$ yields 6.167 volts. Thevenin's resistance ($R_{Thev} = R_{eq}$) will now be found by using the simpler method of merely replacing all independent voltage sources with zero, and all independent current sources with an open circuit; then, looking back into the circuit from *x-x*, find the equivalent resistance. This resistance yields $R_{OC} = 1.667\Omega$. From the maximum power relationship, $R_L = R_{eq}$. The power dissipated is $R_L = 5.71$ watts (see Fig. 1.2).

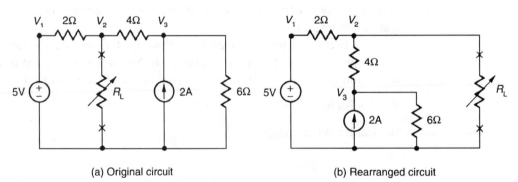

(a) Original circuit   (b) Rearranged circuit

Figure 1.1 Sample problem for maximum power transfer

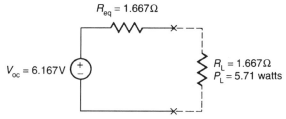

Figure 1.2   Final load resistance using Thevenin's circuit

For a somewhat more difficult problem, a dependent source will now be introduced. Again, consider the circuit of Figure 1.1: if the circuit were modified such that either of the independent voltage or current sources were replaced with a dependent one (see Fig. 1.3a), Thevenin's voltage is calculated and found to be 4.33 V. This value is found by summing the currents at voltage node $V_2(V_{OC})$ as before, and at V_3 (where $I_D = 0.5I_1 = -(V_2 - 5)/2$), as

$$\Sigma\text{I's @ } V_3: \quad \frac{(V_3 - V_2)}{4} + \frac{V_3}{6} - I_D = 0$$

Then, to find Thevenin's resistance, you can determine the short circuit current and find the resistance in the same manner as before. (Here, you could use a voltage source and a series resistance in place of a current source parallel resistance dual for the dependent current source.) The short circuit current is found to be 3.25 A; then Thevenin's resistance is $R_{eq} = V_{OC}/I_{SC} = 1.33$ ohms.

Another method of finding R_{eq} is to remove all independent sources and replace them with their correct resistances, but this approach requires caution. Here, the dependent source would remain in the circuit when finding Thevenin's resistance. To have a dependent source function properly, there must be a current or voltage to be dependent on; because all independent sources have been temporarily removed from the circuit, it is necessary to provisionally insert a fictitious voltage or current into the circuit. This is done by applying the temporary voltage (or current) at the terminals x-x (see Fig. 1.4). A good value to insert is the voltage already calculated to be Thevenin's voltage. Then solve for the current through this temporary voltage as a result of this insertion. Thevenin's resistance will then be V_{OC}/I_{Temp}. The current I_1 may be found directly (the original voltage is zero when making this calculation), $I_1 = -2.17$ A; this immediately yields the dependent current source value of -1.08 A.

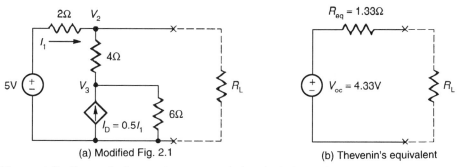

(a) Modified Fig. 2.1   (b) Thevenin's equivalent

Figure 1.3   Dependent current source replacing the independent source of Figure 1.1b

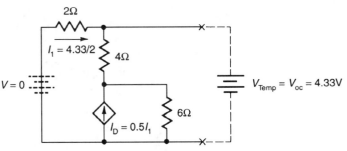

Figure 1.4   Independent source(s) removed

This allows you to set up a node voltage relationship to calculate an I_{Temp} of 3.25A (2.17 + 1.08). Thevenin's resistance is calculated to be $R_t = 4.33/3.25 = 1.33$ ohms.

TRANSIENT ANALYSIS

Transient analysis problems that involve "dc circuits" are usually in steady state conditions, and at some arbitrary time a switch in the circuit is either opened or closed, changing the steady state relationship. To find the steady state conditions prior to the opening or closing of a switch, imagine all capacitors to be open circuits and inductors to be short circuits. Compute the voltage across these open circuits and the currents through these short circuits. The values found this way can be used as initial conditions for the circuits at the 0^+ time (immediately after the switch is changed).

Exceptions to this rule are found when an impulse of voltage or current occurs. An example would be connecting two capacitors that have two different voltage charges together in parallel with no series resistance between them. This kind of problem can be handled by a "conservation of charge" principle. That is, the total charge on the two capacitors must be the same before and after the connection. A similar condition is true for two series inductors with no shunt resistances. In this case, a "conservation of flux linkages" must hold.

Transient solutions, for systems involving a single loop or node analysis, are generally solved most easily by writing a linear differential equation and solving by classical methods.

As an example, assume that a capacitor of 4 μf with no charge on it is connected through a resistance of 15 kΩ to a dc network with voltage source of 80 volts through a switch as shown in Figure 1.5. Of course the network is merely

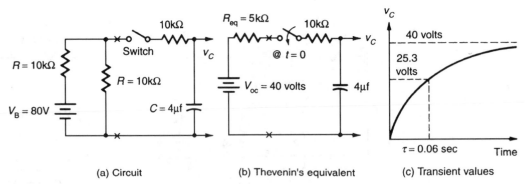

(a) Circuit   (b) Thevenin's equivalent   (c) Transient values

Figure 1.5   A first order R-C transient system

reduced to its Thevenin equivalent to produce a simple series circuit voltage on the capacitor, building up exponentially to the source voltage of 40 volts while the current continually decreases to zero. For this circuit, it is obvious that the RC time constant is 0.06 seconds. Of course, if you are interested in the voltage at any particular time, say at its time constant value, then substituting this value gives

$$V_C(t = 0.06) = 40(1 - e^{-1}) = 40(63.2\%) = 25.3 \text{ volts.}$$

For a second order system (say two energy storage elements in a loop), one again follows the technique of writing the describing differential equation for the loop. Then all that is necessary is to determine the characteristic equation, the natural frequencies, and the correct form of the natural and forced response. Consider the following differential equation:

$$3d^2i/di^2 + 12 \ di/dt + 24i = 3u(t), t > 0,$$

(Usually we normalize the coefficient of the highest derivative term—this allows for easy use of standardized second order curves as shown in Chapter 6.)

$$d^2i/di^2 + 4 \ di/dt + 8i = 1u(t), t > 0.$$

The characteristic equation is easily found to be

$$\alpha^2 + 4\alpha + 8 = 0, \ \alpha = -2 \pm (1/2)\sqrt{16 - 4 \times 8} = -2 \pm j2$$

and the natural response form gives

$$i_n(t) = k_1 e^{-(2+j2)t} + k_2 e^{-(2-j2)t}$$

The forced response is of the form

$$i_f(t) = k_3.$$

Of course the total solution is the sum of the natural and forced response with the constants evaluated for the particular initial conditions.

For a system that involves several loops or nodes, a system of equations using the Laplace variable "s" should be written and solved for by matrix methods. Recognizing that the solution will be the form Ke^{st}, a term for *each* value of s is found from the characteristic equation (the denominator polynomial is found from the matrix solution of the network equations).

IMPEDANCE MATCHING

Impedance matching in circuits for maximum power transfer usually involves finding the Thevenin's impedance of a circuit and making the load the complex conjugate of this impedance. However, the calculation sometimes can be troublesome if an included dependent source is involved. Finding Thevenin's open circuit voltage is straightforward, but recall that there are two ways used to calculate Thevenin's impedance. One approach is to replace the independent sources with their internal impedances and then find the impedance looking back into the circuit. However, this creates a problem with what to do with the dependent sources. As previously discussed, by applying a temporary external source at the terminals used for Thevenin's equivalent circuit, a calculation of the value of the dependent source may be made. If you choose to use Thevenin's voltage for this value of the provisional source, the resulting current through this source allows you to calculate Thevenin's impedance directly, $Z_{eq} = V_{OC}/I_{Temp}$.

Another way of solving for Thevenin's impedance is to merely short the active circuit at terminals x-x, and determine the short circuit current. With the original dependent sources still in place, the dependent sources may be calculated. The equivalent impedance is then found, $Z_{eq} = V_{OC}/I_{SC} = Z_{Thev}$.

This latter technique will be demonstrated in the following steady state ac network example with a dependent current source (which depends on a developed voltage across a circuit element). Assume it is required to find Z_L for maximum power transfer for the external load at terminals x-x (see Fig. 1.6a). Here you can use either the jX_L and $-jX_C$ notation, or use the Laplace operator "s" and eventually let $s \rightarrow j\omega$ since the circuit is operating in steady state as in Figure 1.6b. Assume the source voltage $v_g = 100\cos(5{,}000t)$ and is the reference quantity.

Summing current at various nodes gives

$$\Sigma\text{I's at node } V_1: \quad \frac{(V_1 - V_g)}{\left[\frac{1}{(sC)}\right]} - (0.02V_x) + \frac{(V_1 - V_x)}{sL} = 0$$

$$sLV_1 - SLV_g + \frac{(sL)}{sC}[(-0.02V_x)] + \left[\frac{1}{(sC)}\right]V_1 - \left[\frac{1}{(sC)}\right]V_x = 0$$

$$(s^2LC)V_1 - (s^2LC)V_g - (sL)(0.02V_x) + V_1 - V_x = 0$$

$$(s^2LC + 1)V_1 - (1 + 0.02sL)V_x = s^2LCV_g$$

$$\Sigma\text{I's at node } V_x:$$

$$\frac{(V_x - V_1)}{sL} + \frac{V_x}{R} = 0; \quad RV_x - RV_1 + sLV_x = 0 \quad \boxed{V_1 = \left(1 + \frac{sL}{R}\right)V_x}$$

$$(R - SL)V_x - RV_1 = 0$$

Combining nodal equations, $(s^2LC + 1)\left(1 + \dfrac{sL}{R}\right)V_x - (1 + 0.02sL)V_x = s^2LCV_g$.

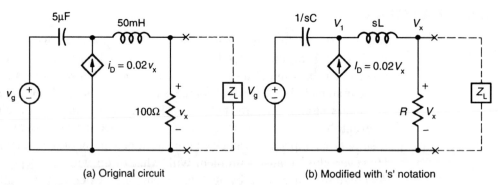

(a) Original circuit   (b) Modified with 's' notation

Figure 1.6   Circuit with dependent source

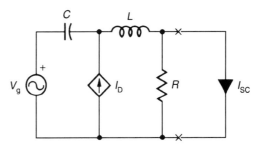

Figure 1.7 Shorted load circuit

Because the circuit is in steady state, let $s \rightarrow j\omega = j5 \times 10^{3}$; and $V_{g} = 100/2$ (for rms values).

$$(-25 \times 10^{6} \times 50 \times 10^{-3} \times 5 \times 10^{-6} + 1)\left(1 + \frac{j5 \times 10^{3} \times 50 \times 10^{-3}}{100}\right)V_{x} +$$

$$-(1 + 0.02j5 \times 10^{3} \times 50 \times 10^{-3})V_{x} = -25 \times 10^{6} \times 50 \times 10^{-3} \times 5 \times 10^{-6}V_{g}$$

$$(-6.25 - j18.1)V_{x} = -6.25V_{g} = -6.25 \times \frac{100}{\sqrt{2}}\angle 0°; \quad 19.2\angle 71°V_{x} = 442\angle 0°$$

$$\left|V_{x}\right| = \frac{442\angle 0°}{19.2\angle 71°} = 23.1\text{V}$$

To find I_{SC} place a short circuit across terminals x-x and calculate the current through this short circuit (see Fig. 1.7).

Then sum the current at the node V_{1}.

$$\Sigma \text{ I's at } V_{1}: \quad \frac{(V_{1} - V_{s})}{\frac{1}{(sC)}} + \frac{(V_{1} - 0)}{sL} = 0; \quad sLV_{1} - sLV_{g} + \left(\frac{1}{sC}\right)V_{1} = 0$$

$$sLCV_{1} - s^{2}LCV_{g} + V_{1} = 0; \quad (s^{2}LC + 1)V_{1} = s^{2}LCV_{g}$$

Again, let $s \rightarrow j\omega = j5 \times 10^{3}$.

$$(-6.25 + 1)V_{1} = -6.25 \times 100\angle 0°$$

$$-5.25V_{1} = 625\angle 0°$$

$$V_{1} = \left(\frac{625}{5.25}\right) = 119\angle 0°$$

$$I_{sc} = I_{x} = \frac{V_{1}}{(sL)} = \frac{119\angle 0°}{j250} = 0.476\angle -90°$$

$$Z_{eq} = \frac{V_{oc}}{I_{sc}} = \frac{(32.6\angle -71°)}{(0.476\angle -90°)} = 68.5\angle 19°$$

$$Z_{Thev} = Z_{eq} = 68.5\angle 19° = 64.8 + j22.3$$

From Figure 1.8, using Z_{L} = complex conjugate of Z_{eq}, the real power in the load is easily calculated to be

$$I = \frac{V_{oc}}{(Z_{eq} + Z_{L})} = \frac{(23.1\angle -71°)}{(64.8 + j22.3 + 64.8 - j22.3)} = 0.178\angle -71°$$

$$P_{L} = I^{2}R_{L} = (0.178)^{2}64.8 = 2.06 \text{ watts.}$$

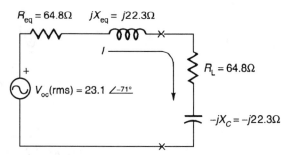

Figure 1.8   Thevenin's equivalent circuit with a complex conjugate load

Another form of impedance matching involves the use of transformers. Consider an ideal lossless transformer used to match a load to an ac source with turns ratio, "a"; here $a = N_1/N_2 = |V_p/V_S| = |V_1/V_2| = |I_S/I_P| = |I_2/I_1|$. Then the impedance of the load side as seen on the primary side is $Z_P = a^2 Z_s$.

As a short example, assume a voltage source is 50 volts with an internal impedance of $100 + j50$ ohms, while that of a load is $10 - j5$ ohms (see Fig. 1.9). If connected together without the transformer, the power dissipated in the source may be found from $I = 50\angle 0°/(100 + j50 + 10 - j5) = 0.421/-22.4°$, giving $P_{\text{Source}} = I^2 \times 100 = 17.7$ watts while that of load is $P_{\text{Load}} = I^2 \times 10 = 1.77$ watts! On the other hand, by using a transformer with an $a = /10$, the load as seen on the source side is $a^2 Z_L$ or $(/10)^2 (10 - j5) = 100 - j50$. The power dissipated in the source and load respectively are now found from $I = 50\angle 0°/(100 + j50 + 100 - j50) = 0.25 \angle 0°$, $P_{\text{Source}} = P_{\text{Load}} = I^2 R = 6.25$ watts each.

A somewhat more comprehensive situation is involved for a nonideal transformer (see Chapter 2), or for an air core transformer with a mutual inductance, M, whose coefficient of coupling, k, is less than unity. Recall that $M = k/L_1 L_2$ and that the dot notation is used to show relative polarity. Also recall the reflected impedance (of the secondary reflected to the primary), Z_r, is no longer just $a^2 Z_L$ but now is a function of k, ω, L_1, L_2, and the load impedance. The following example will illustrate the calculations involved for k less than one and for real values of L's.

Assume an ac voltage source with an internal resistance is coupled to a load impedance, Z_L, by two air core coils with a coupling coefficient of 0.5. Determine the power dissipated in the load and whether load is matched to the source (see Fig. 1.10).

The mutual inductance is given by $M = k\sqrt{L_1 L_2}$

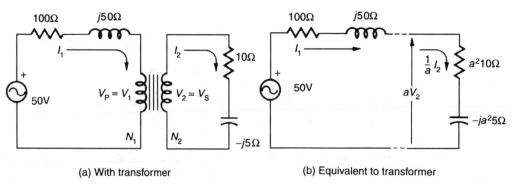

(a) With transformer   (b) Equivalent to transformer

Figure 1.9   Simple ac circuit with transformer and equivalent

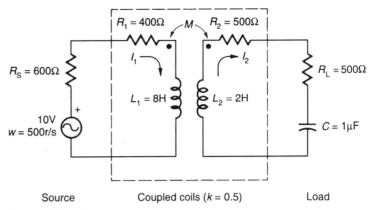

Source   Coupled coils ($k = 0.5$)   Load

Figure 1.10   Example coupled circuit

henrys and the mutual reactance by $jX_M = j\omega M$ ohms.

$$M = 0.5\sqrt{8\times2} = 2H; \quad jX_M = j500\times2 = j1,000\Omega$$

$$jX_{L1} = j\omega L_1 = j500\times8 = j4000\Omega; \quad jX_{L2} = j500\times2 = j1,000\Omega$$

$$-jX_C = -j\frac{1}{\omega C} = -j\frac{1}{500\times1\times10^{-6}} = -j2000\Omega$$

$$\mathbf{Z(Primary)} = Z_{11} = 600 + 400 + j4,000 = (1+j4)\times10^3\Omega$$

$$\mathbf{Z(Secondary)} = Z_{22} = 500 + 500 + j1,000 - j2,000 = (1-j1)10^3\Omega$$

The reflected impedance, Z_r (from the load side to the primary side), is given by something similar to "a^2," but is actually $X^2_M/|Z_{22}|^2$ and is then multiplied by the conjugate of Z_2,

$$Z_r = \left|\frac{X_M}{Z_{22}}\right|^2 Z_{22}(conj) = \left|\frac{10^3}{\sqrt{2}\times10^3}\right|^2 (1+j1)\times10^3 = (0.5+j0.5)\times10^3$$

$$= \sqrt{2}\times0.5\times10^3\angle45°$$

The currents may be found directly as can the power taken from the source and the power dissipated in the load.

$$I_1 = \frac{V\angle0°}{Z_{11}+Z_r} = \frac{10\angle0°}{(1+j4+0.5+j0.5)\times10^3} = \frac{10\angle0°}{(1.5+j4.5)10^3} = 2.11\angle-71.6° \text{ mA}$$

$$I_2 = \frac{jX_m}{Z_{22}}I_1 = \left(\frac{10^3\angle90°}{\sqrt{2}\times10^3\angle-45°}\right)(2.11\angle-71.6°) = 1.5\angle-26.6° \text{ mA}$$

We would like the efficiency of the circuit powers to approach 50 percent for maximum power transfer (*i.e.*, $P_L/P_{Source} \to 1.0$), but of course, this can never be achieved because of the real parameter of the coils and that of k being less than unity.

$$P_L = I_2^2 R_L = (1.5)^2\times10^{-6}\times500 = 1.12\times10^{-3} \text{ watts}$$

$$P_S = VI\cos\phi = (10\times2.11)\cos71.6°\times10^{-3} = 6.67\times10^{-3} \text{ watts}$$

$$\boldsymbol{Eff} = \frac{P_L}{P_S} = \frac{1.12\times10^{-3}}{6.67\times10^{-3}} = 16.8\%$$

When mutual coupling is involved with k less than unity, impedance matching becomes difficult. However, if the real parts of both the source and load impedance cannot be altered, the series reactance of the load can usually be changed to cancel the source reactance. The effect is to maximize the current and thus maximize the power in the load. For more reading on this subject, refer to the Nilsson and Bobrow texts listed in the references at the end of this chapter.

Finally, other forms of impedance matching are involved with transmission lines. One form is at lower frequencies (power lines) and another at higher frequencies (open wire pair, coaxial cables, wave guides, etc.). Power line transmission is covered in Chapter 3 while transmission lines, in general, are presented in Chapter 8.

METERS AND WAVEFORMS

With newer type meters, such as digital multimeters (DMM), even the simplest of measurements may cause a reading to be incorrectly interrupted. Of course the older analog type meters presented the same kinds of problems unless one knew something about the waveform being measured. This short review will address both the kind of meters being used and waveforms being measured, along with various kinds of notation.

Before the advent of digital electronics, the basic tool for measuring dc currents was the D'Arsonval meter. It is important to review certain aspects of measurement using this kind of meter. Recall that current flow in the presence of a magnetic field produces a force ($f = Bil$). The basic design of the meter entails arranging a permanent magnet such that an air gap is in a radial form, with wire formed in a rectangular coil with an indicating needle attached to it along with a restraining spiral spring (see Fig. 1.11).

The torque produced is proportional to current if all other parameters are fixed ($T = 2N\,Blri = K_1 i$). This developed torque will just equal the restraining torque of the spiral springs (torque = $K_s \Theta$); the angular displacement of the needle is easily found to be proportional to the current, $\Theta = K_\Theta i$.

Because full scale on the D'Arsonval meter may represent a very small current, the coil is usually shunted with a very low resistance when measuring currents. This meter may be used to measure voltage as well; this is done by putting a high resistance in series with the coil. This meter will read the average value of current through the coil. Refer to Figure 1.12.

For a short example using the D'Arsonval meter, assume 10 ma gives full-scale reading and the coil resistance is 5.0 ohms. What series resistance should

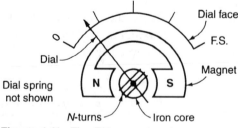

Figure 1.11 The D'Arsonval meter movement

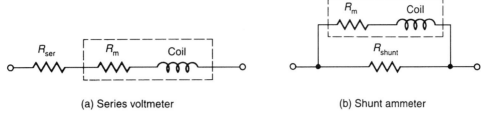

(a) Series voltmeter   (b) Shunt ammeter

Figure 1.12   A D'Arsonval voltmeter/ammeter circuit

be added in series so it will act as a voltmeter with a 150-volt range? The resistor size is found from

$$R_{tot} = V/I = 150/(10 \times 10^{-3}) = 15,000 \ \Omega$$
$$R_{ser} = 15,000 - 5 = 14,995 \ \Omega$$
$$(\text{Or, } R_{ser} = 15,000 \text{ with only } 0.033\% \text{ error.})$$

For the D'Arsonval meter to act as ammeter with full-scale deflection of 20 A, what shunt resistor is needed in parallel for 10ma to flow through the meter? The voltage across the meter is $V_m = IR = (10 \times 10^{-3})5.0 = 50$ mv, so the voltage across the shunt must be the same. Because for full-scale deflection, the meter takes 10 ma, the shunt must pass 19.99 amperes, therefore

$$R_{shunt} = V/I = (50 \times 10^{-3})/19.99 = 2.501 \times 10^{-3}\Omega.$$

An extension to the problem might be to use the D'Arsonval meter to measure an ac voltage (here we assume the inductance of the coil is negligible). Of course, the meter with the series resistor would read the average voltage, which is zero for a sinusoidal voltage. However, using an ideal diode in series with R_{ser}, the full-scale rms voltage (see Fig. 1.13) may be found.

The average value for a clipped sinusoidal signal (a half sine wave spread over a full period (see Table 1.1) is found to be $V_{avg} = V_m/\pi$) and should equal the average voltage of the previously described 150 volt dc meter. Thus the supplied ac voltage must have a peak (maximum) voltage of $V_m = \pi V_{avg} = \pi 150 = 471$ volts and the applied rms voltage is $V_{rms} = V_m/2 = 471/1.41 = 333$ volts.

In contrast to the D'Arsonval type meter, the DMM of today requires a little more knowledge of both the wave form and the meter itself. Many such meters (as an example, the Fluke DMM, 8010 series) will measure the true rms voltage of an ac signal. However, if there is a dc component, it will have to be measured

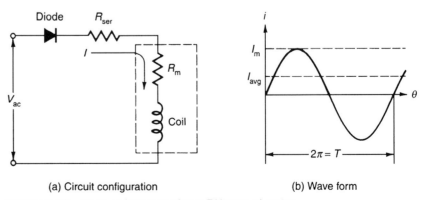

(a) Circuit configuration   (b) Wave form

Figure 1.13   An ac voltmeter using a D'Arsonval meter

Table 1.1 Average and rms values for various waveforms

Sinusoidal	Square	Triangular
$I_{avg}=0$ $I_{rms}=I_{max}/\sqrt{2}$	$I_{avg}=0$ $I_{rms}=I_{max}$	$I_{avg}=0$ $I_{rms}=I_{max}/\sqrt{3}$
$I_{avg}=I_{max}/\pi$ $I_{rms}=I_{max}/2$	$I_{avg}=I_{max}/2$ $I_{rms}=I_{max}/\sqrt{2}$	$I_{avg}=I_{max}/4$ $I_{rms}=I_{max}/\sqrt{6}$
$I_{avg}=2I_{max}/\pi$ $I_{rms}=I_{max}/\sqrt{2}$	$I_{avg}=I_{max}$ $I_{rms}=I_{max}$	$I_{avg}=I_{max}/2$ $I_{rms}=I_{max}/\sqrt{3}$

separately; the relationship between the total rms value of the waveform and the ac component and the dc component is given as

$$\text{rms (total)} = \sqrt{(\text{ac component, rms})^2 + (\text{dc component})^2} \qquad (1.1)$$

For current measurements, when the DMM is inserted in series with the circuit, you may have to consider the voltage drop across the DMM itself; and if so, the current being measured will be in error. This voltage drop is referred to as the **burden voltage**. These burden voltages vary for different scales of the particular meter. As an example, a typical burden voltage, V_B, might be 0.3 volts for a 200 mA scale and 0.9 volts on the 2,000 mA scale (of course, when measuring full-scale values). The displayed reading as a percentage of full scale will be [(100 × actual reading)/(full scale)] × (full scale burden voltage). As an example, suppose the current is being measured in a simple series circuit made up of a 10 volt source and load resistance of 7.5 ohms; the current expected is $I = V/R = 10/7.5 = 1.33$ A. However, the displayed current is, say, 1.254 A; this takes into account the burden voltage (0.9 volts on the 2,000 mA scale), $V_B = 100 \times (1.254/2.000) = 62.7\%$ of $0.9 = 0.564$ volts. The percentage error is $100 \times 0.564/(10 - 0.564) = 5.98\%$. To obtain the true current, increase the displayed current by the 5.98% error: $I = 1.254 \times (1 + 0.0598) = 1.33$ A.

When making or calculating measurements, you must always keep in mind the basic fundamentals with regard to wave forms and the limits of the measuring device.

Recall that when determining the average value of a quantity, say a current, this value would be the average value of a varying current over a period of time that would deliver the same charge as a constant current over the same period. That is,

$$I_{avg}T = Q = \int_0^T i(t)dt \quad I_{avg} = \frac{1}{T}\int_0^T i(t)dt. \tag{1.2}$$

And, for an rms value (again, say, current), the classical value is defined as being numerically equivalent to providing the same heating effect through a resistance as a constant dc source would over the same period of time. That is, the average value of a varying power $p(t)$ over the same time period would transfer the same energy, W. Thus (note that P_{avg} is always the same meaning as P without a subscript),

$$P_{avg}T = W = \int_0^T p(t)dt \quad P = \frac{1}{T}\int_0^T p(t)dt = \frac{1}{T}\int_0^T i^2R = I^2R = I_{rms}^2R \tag{1.3}$$

which yields the standard equation for the effective or rms values of a periodic wave,

$$X_{rms} = \sqrt{\left(\frac{1}{T}\right)\int_0^T x^2(t)dt}. \tag{1.4}$$

A few of the standard waveforms with their average and rms values are given in Table 1.1.

If the waveform is periodic, both the average and rms values are relatively easy to determine. Consider the voltage waveform shown in Figure 1.14. Because the waveforms are rectangular in shape, both the average and rms values are found by realizing what the previous equation really means. For this figure, the period, T, is seen to be 4 seconds and the average and rms values are given by

$$V_{ag} = \text{(Net Area Over One Period)/(Time of Period)}$$
$$= (1 \times 1 + 3 \times 2 + 0 \times 1)/4 = 1.75 \text{ V}$$
$$V_{rms} = (\Sigma \text{ of Areas of Voltage Squared Over One}$$
$$\text{Period)/(Time of One Period)}$$

$$V_{rms} = \sqrt{\frac{1^2 \times 1 + 3^2 \times 2 + 0^2 \times 1}{4}} = 2.18 \text{ V}.$$

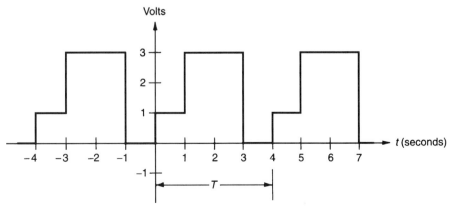

Figure 1.14 A periodic waveform

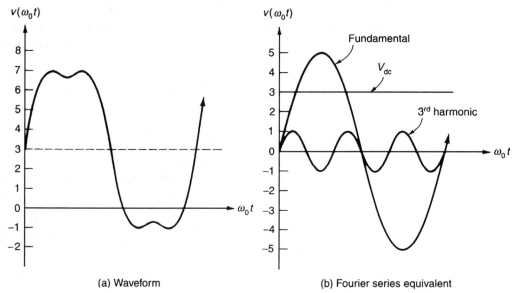

Figure 1.15   A distorted wave with dc level

Also of interest in waveform measurements are their Fourier coefficients. As an example, the rms value of a Fourier series is the same as that found by using the standard equation, as above, except the representation for the function squared is replaced with the series

$$F_{rms} = \sqrt{(1/T)[a_{dc} + \Sigma A_n \cos(n\omega_0 - \Theta_n)]^2 \, dt}. \tag{1.5a}$$

Because the function is squared and is integrated over one period, it is easily shown that the equation reduces to

$$F_{rms} = \sqrt{(1/T)\left[(a_{dc})^2 T + \Sigma(T/2)A_n^2 \right]} = \sqrt{a_{dc}^2 + \Sigma(A_n/\sqrt{2})^2}. \tag{1.5b}$$

This equation may now be explained in words to mean the square root of the sum of the dc term squared plus the square of all the rms values of the harmonics. As an example, suppose a voltage waveform, from the output of a slightly saturated transformer in series with a 3-volt battery, is as given in Figure 1.15. The equation is given by a dc value of 3 volts, and a fundamental wave of $5\cos(\omega_o t + \Theta_1)$ and a third harmonic of $\cos(3\omega_o t + \Theta_3)$ volts, and the rms value of the voltage is

$$V_{rms} = \sqrt{3^2 + (5/\sqrt{2})^2 + (1/\sqrt{2})^2} = \sqrt{22} = 4.69 \text{ V}.$$

RESONANCE AND BANDWIDTH CALCULATIONS

The basic relationship of resonance deals with energy storage. If no power-consuming element is involved (*i.e.,* no resistance), then energy storage amounts to exchanging energy between a capacitor and inductor with no loss of energy.

Recall that the energy stored in a capacitor is $w_C = (1/2)Cv^2(t)$ and for an inductor is $w_L = (1/2)Li^2(t)$. For a series of steady state ac circuits in resonance (that may include a resistive element), the capacitive and the inductive reactance just cancel to make the impedance appear as purely resistive (if any). The energy stored in either element at any instant, of course, may be found by the previous equations.

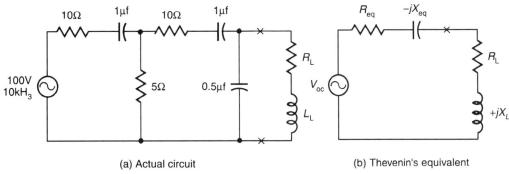

(a) Actual circuit　　　　　　(b) Thevenin's equivalent

Figure 1.16　Circuit reduction for resonance

If a circuits problem involving series resonance seems unclear as to method of solution, one approach is to isolate all of either the inductors or capacitors to one side of a circuit and use either Thevenin's or Norton's theorem for the opposite element.

A short example will demonstrate the method. Assume it is desired to find the size of the inductor to give maximum voltage across its terminals (see Fig. 1.16).

Because the frequency is given as 10 kHz, X_{C1} and X_{C2} are easily found to be 15.9 and 31.8 ohms, respectively. Setting and solving the loop equations will yield V_2 directly, which is Thevenin's voltage, V_{oc}:

Loop 1:　$100\angle 0° = (10 - j15.9)\,I_1 + 5(I_1 - I_2) = (15 - j15.9)I_1 + (-5)I_2$

Loop 2:　$0 = (10 - j15.9)I_2 - j31.8I_2 + 5(I_2 - I_1) = (-5)I_1 + (15 - j47.8)I_2,$

$$I_2 = \frac{\begin{vmatrix} 15 - j15.9 & 100\angle 0° \\ -5 & 0 \end{vmatrix}}{\begin{vmatrix} 15 - j15.9 & -5 \\ -5 & 15 - j47.8 \end{vmatrix}} = \frac{0 + 500\angle 0°}{1096\angle -119.5° - 25\angle 0°} = 0.451\angle 120.4° \text{ A}$$

$V_2 = V_{OC} = I_2(-j31.8) = (0.451\angle 120.4)\,(-31.8\angle -90°) = 14.3\angle 30.4° \text{ A}.$

Then, shorting the terminals at x-x and solving for this short circuit current, I_{SC} will allow you to obtain Thevenin's resistance, R_q:

Loop 1:　$100\angle 0° = [(10 + 5) - j15.9]I_1 + (-5)I_2$

Loop 2:　$0 = 5(I_2 - I_1) + (10 - j15.9)I_2 = (-5)I_1 + (15 - j15.9)I_2$

$$I_2 = I_{SC} = \frac{\begin{vmatrix} 15 - j15.9 & 100\angle 0° \\ -5 & 0 \end{vmatrix}}{\begin{vmatrix} 15 - j15.9 & -5 \\ -5 & 15 - j15.9 \end{vmatrix}} = \frac{0 + 500\angle 0°}{478\angle -93.4° - 25\angle 0°} = 1.04\angle 96.4° \text{ A}$$

$Z_{eq} = V_{OC}/I_{SC} = 14.4\angle 30.4°/1.04\angle 96.4° = 13.8\angle -66° = 5.61 - j12.6 \ \Omega$

thus, $X_L = 12.6\Omega$, $L = 12.6/(2\pi f) = .201$ mH.

For maximum power transfer, Z_L should be the complex conjugate of Z_{eq}, but the question was for maximum developed voltage across the inductor. Therefore you should select the inductance for being as close to 0.2 mH and with as low internal resistance as possible for the highest possible resonant peak. If the frequency is controllable and can be moved in either direction from resonance and the values of the reactances are near that of the resistance, the maximum current reduces slowly with frequency. When the resistance is much smaller than the reactances, the current decreases rapidly away from resonance.

Series Resonance

For a series circuit, the ratio of either the inductive or capacitive reactance to resistance is called the Q (quality factor) of the circuit.

Since $X_C = X_L$, $\omega_r = \dfrac{1}{\sqrt{LC}}$,

Q is given by

$$Q = \omega_r L/R = 1/(\omega_r RC) = (1/R) \sqrt{L/C} \qquad (1.6)$$

and the bandwidth (BW $= \omega_2 - \omega_1$) is

$$BW = \omega_r/Q = R/L = 1/(RC) \qquad (1.7)$$

where the half power frequency points, ω_1 and ω_2, may be found in terms of Q as

$$\omega_1 = \omega_r[\sqrt{[1/(2Q)]^2 + 1} - 1/(2Q)] \qquad (1.8a)$$

$$\omega_2 = \omega_r[\sqrt{[1/(2Q)]^2 + 1} + 1/(2Q)] \qquad (1.8b)$$

For a $Q \gg 10$, equations 1.8a and 1.8b reduce to

$$\omega_{1,2} = \omega_r[1 -/+ 1/(2Q)], \quad \text{for} \quad Q \gg 10. \qquad (1.8c)$$

As an example, if $R = 2\Omega$, $L = 5h$, $C = 400 \ \mu F$, the resonant frequency, the Q, and the bandwidth are easily found to be

$$\omega_r = 1/(\sqrt{LC}) = 1/(\sqrt{5 \times 400 \times 10^{-6}}) = 22.36 \text{ r/s}$$

$$Q = (1/R)(\sqrt{LC}) = (1/2)(\sqrt{5 \times 400 \times 10^{-6}}) = 55.9$$

$$\omega_{1,2} = 22.36\left[1 \pm \frac{1}{(2 \times 55.9)}\right] = 22.36 \pm 0.2 = 22.16, \quad 22.56 \text{ r/s},$$

$$BW = \omega_2 - \omega_1 = 22.56 - 22.16 = 0.4 \text{ r/s}$$

Check: $\qquad\qquad BW = R/L = 2/5 = 0.4 \text{ r/s}.$

Parallel Resonance

For parallel circuits that have pure R, L, and C elements (see Fig 1.17a), the total current at resonant frequency will be a minimum as both capacitive and inductive currents are equal but of opposite phase and will cancel ($Y = G + jB_C - jB_L$). The total current to the parallel branch will be as though only the resistance is in the circuit; however at frequencies below resonance the current through L will be larger than through C, increasing the net current.

Again, for a pure RLC parallel circuit, the resonant frequency, ω_r, is the same as a series resonant circuit

$$(\omega_r = 1/\sqrt{LC})$$

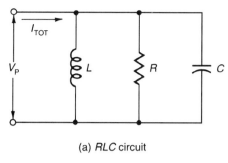 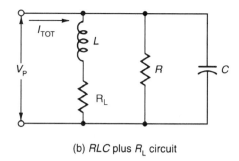

(a) *RLC* circuit   (b) *RLC* plus R_L circuit

Figure 1.17   An *RLC* parallel circuit

while for Q, it is given as

$$Q = \omega_r RC = R/(\omega_r L) = R\sqrt{C/L}. \tag{1.9a}$$

An alternate definition of Q is the ratio of 2π times the energy stored per cycle to the energy dissipated per cycle (see Eq. 1.9c.) Thus,

$$Q = 2\pi[\omega_C(t) + \omega_L(t)]/[W_R] \tag{1.9b}$$

where $W_R = P_R T$ (*i.e.*, average power loss x period/cycle). Thus,

$$Q = \frac{2\pi 0.5 CR^2 I^2}{\pi I^2 R/\omega_r} = \frac{2\pi(0.5 R^2 I^2)}{(\pi I^2 R\omega_r L)}$$

$$= \omega_r RC = R/(\omega_r L) = R\sqrt{C/L} \tag{1.9c}$$

For the bandwidth, equations for ω_1 and ω_2 are the same as for series resonance (see Eqs 1.8a,b) if the parallel circuit is pure (see Fig. 1.17a). However, for a typical parallel resonant circuit, there is often some resistance that is associated with the inductance (see Fig. 1.17b); usually the coil resistance in series with the inductance tends to slightly increase the complexity of computation at resonance.

The admittance equation for Figure 1.17b now becomes

$$Y = 1/R + 1/(R_L + j\omega L) + j\omega C = 1/R + (R_L - j\omega L)/(R_L 2 + \omega^2 L^2) + j\omega C. \tag{1.10}$$

The resonant frequency, ω_r, is when the *j*-terms cancel, making the circuit purely conductive. Solving Equation 1.10 for this cancellation, ω_r becomes

$$\omega_r = \sqrt{1/(LC) - (R_L/L)^2}. \tag{1.11}$$

It should be noted that if $R_L/L \rightarrow 0$, then ω_r is $1/\sqrt{(LC)}$, the same as series resonance. The value of the admittance at ω_r becomes

$$Y(\omega_r) = 1/R + R_L C/L = (L + RR_L)/(LR) \tag{1.12}$$

If the circuit (of Fig. 1.17b) is connected to a constant current source, the amplitude of voltage output is given by

$$V_o = [RL/(RR_L C + L)] (I_{source}) \tag{1.13}$$

But Equation 1.13 does not necessarily give the maximum output voltage. You can develop an equation (a lengthy process) for the frequency for maximum output voltage. The result of the development would yield a frequency, ω_m, to give maximum voltage,

$$\omega_m = \sqrt{\sqrt{[(1/(LC)^2(1+2R_L/R)+(R_L/L)^2(2/(LC)]-(R_L/L)^2}} \qquad (1.14)$$

Using this new frequency with the admittance equations allows you to calculate the maximum output voltage at ω_m; of course, if $R_L \rightarrow 0$, then $\omega_m \rightarrow \omega_r$.

More information on resonance and method of solution are presented in the Chapter 5 on control systems. These other methods are based on both root locus techniques and Bode plot analysis.

ELECTROSTATICS AND MAGNETIC FIELDS

Only an introductory level of both electric and magnetic fields is reviewed here. Subsequent chapters present more comprehensive aspects.

Electric Fields

Usually you especially need to review electric fields and flux densities due to electrical charges. These fields, forces, and flux densities may require three-dimensional vector notation. However, when problems reduce to two-dimensional ones, a "quick" graphical solution may give a faster answer (particularly true for the multiple choice-type answers). For notational purposes, use the notation with which you are already familiar.

Our review starts with stationary electric charges that produce electric fields. These fields may be defined in terms of the forces they produce on one another. The smallest amount of charge that can exist is the charge of one electron, which is 1.602×10^{-19} coulombs (C). One coulomb of charge is thus equivalent of 6.24×10^{18} electrons. This is the amount of charge that is necessary to develop a force of 1 newton in an electric field of 1 volt per meter. An electric field, E (bold face for vector quantity), is nothing more than the amount of force (f) that would be exerted on a positive charge (assuming it is concentrated at a point) of one coulomb if placed in that field, as follows:

$$F = QE. \qquad (1.15)$$

The electric field is not thought of as a point, but rather as being distributed throughout a small region. This is a vector quantity; the direction of force on the one coulomb would be toward the point source of the field if that point were a negative charge. The units of measurement are newtons per unit of charge; alternatively, it may be given in terms of volts/meter because

force/charge = (force × distance)/(charge × distance)
= energy/(charge × distance) = voltage/distance.

For point charges, the force is proportional to the product of the two charges and inversely proportional with the square of the distance between them (similar to the laws of gravity). The constant of proportionality depends on the medium between them. This constant is $1/(4\pi)$, or approximately 9×10^9 for free space (if not free space, the 9×10^9 is merely divided by the relative permittivity),

$$F = (9 \times 10^9/\varepsilon_{rel})Q_1Q_2/r^2 \text{ newtons} \qquad (1.16)$$

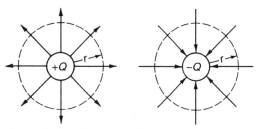

Figure 1.18 Electric flux lines around a charge

where ε is the permittivity of the medium, and

$$\varepsilon = \varepsilon_{rel}\varepsilon_o \ F/m$$

and ε_o = permittivity of free space = 8.85×10^{-12} farads/meter. Permittivity of air is approximately the same as free space or a vacuum (*i.e.*, $\varepsilon_{rel} = 1.0006$; whereas, if the medium happens to be castor oil, ε_{rel} is almost 5).

A convenient way of thinking of these fields is to imagine flux lines radiating either away from or toward a point source (see Fig. 1.18). If you imagine a positive point source charge at the center of a sphere and a number of arrows pointed away from the center, these arrows would be the flux lines (the bigger the charge, the more arrows). The flux density on the surface of the sphere (whose center is located at the point charge) would then be the number of arrows through a unit area on the sphere.

$$D = Q/A, \tag{1.17a}$$

$$Q = D \text{ dot } A \tag{1.17b}$$

where A is the area of the sphere ($4\pi r^2$) and Q is the quantity of charge in terms of coulombs (assuming the area is normal to the flux). Equation 1.17b is presented for those preferring the dot and cross-product vector notation (*i.e.*, this notation guarantees the portion of the surface being considered is normal to the flux). If one were to divide up the area of the sphere into very small areas, *da*'s, the sum of areas times the amount of flux through each area would be the amount of charge at the center of the sphere. More formally presented, as the size of small areas approaches zero, and as one integrates over the entire surface area of the sphere, the total charge enclosed is Q. But, by using the dot product notation, one is not limited to a sphere shape with the charge at the center. The formal law (Gauss' law) is then given as Equation 1.18b or 1.18c.

$$Q = \sum D \, da \text{ (for entire surface)} \tag{1.18a}$$

$$Q = \int D \text{ dot } da \tag{1.18b}$$

$$Q/? = \int E \text{ dot } da \tag{1.18c}$$

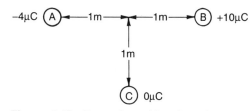

Figure 1.19 Example problem layout

The field strength, **E**, at a sphere surface is then proportional to D,

$$E = (D/\varepsilon)a_r \qquad\qquad (1.19)$$

where a_r is a unit radial vector direction. If there are a number of charges throughout a region, it is usually easier to use the flux density concept for solving problems.

As a short example, assume three point charges, A, B, and C. Points A and B are 2 meters apart; point C is on a perpendicular bisector between A and B and is 1 meter down. Point A has 4×10^{-6} coulombs of negative charge, point B has 10×10^{-6} coulombs of positive charge, and point C has no charge yet. Determine the force and the field on B (in this case, only due to the charge on A) and the flux density at point C.

The force on B due to A is directly proportional to the charges on A and B and inversely proportional to the square of the distance of separation. Using Equation 1.15,

$$F = (9 \times 10^9)\,(4 \times 10^{-6})\,(10 \times 10^{-6})/(2^2) = 9 \times 10^{-2} \text{ newtons}$$

(with the direction of force toward each other).

$$E(\text{at B}) = f/q_B = 9 \times 10^{-2}/10^{-5} = 9 \times 10^3 \text{ V/m}$$

(It is interesting to note that if the medium happened to be a special oil whose relative permittivity is 5.0, both F and E would only be one fifth as large.) A more orderly solution (especially if several charges are involved—or none at C in this case), is to use the flux density relationship to find the individual D's, then convert to E's. All that is now necessary to find the net E is to use vector summation of the flux densities and divide by ε. To find the flux density at C due to charge A is to imagine a sphere passing through C with its center at A; repeat for B.

$$D_{CA}(\text{at C due to A}) = Q_A/(4\pi r^2)$$
$$\mathbf{E}_{CA} = (D/\varepsilon)\,a_{rA} = (Q_A/r^2)(1/(4\pi\varepsilon))a_{rCA}$$

Recall that $1/(4\pi,)$ for free space is 9×10^9, then

$$\mathbf{E}_{CA} = (4 \times 10^{-6})/((/2^2)\,(9 \times 10^9)) = 1.8 \times 10^4 a_{rCA} \text{ V/m}$$
$$\mathbf{E}_{CB} = (10 \times 10^{-6})/((/2^2)\,(9 \times 10^9)) = 4.5 \times 10^4 a_{rCB} \text{ V/m}$$

To find the $\mathbf{E}_{C\,net}$ you may use the more formal procedure of finding the rectangular components of each of the field vectors because only two dimensions are involved. Then, summing the horizontal and vertical components, the net field vector is the square root of the sum of the squares. If we assume the reference vector, a_r, has the horizontal component as "a" and the vertical component as "ja" (the "j" implies the 90° or vertical axis), then

$$\mathbf{E}_{CA} = |\mathbf{E}_{CA}|\,(\cos\Theta + j\sin\Theta) = 1.8 \times 10^4\,(\cos 135° + j\sin 135°)$$
$$= (-1.27 + j\,1.27) \times 10^4 \text{ V/m}$$
$$\mathbf{E}_{CB} = 4.52 \times 10^4\,(\cos 225° + j\sin 225°) = (-3.20 - j3.20) \times 10^4 \text{ V/m}$$
$$\mathbf{E}_{C\,net} = (-4.47 - j1.92) \times 10^4 = 4.87 \times 10^4\angle 203.2° \text{ V/m}$$

where the angle Θ is the angle between the vector E and a horizontal reference line. However, solving for the vector field at C may be done more quickly by using a graphical solution, assuming reasonable accuracy. The points A, B, and C are set up (to their own scale); then, working with point C (almost as a "free

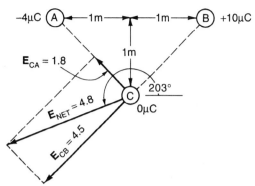

Figure 1.20 Sample problem vector fields
(×10⁴) at C

body diagram"), one places the vectors (using any suitable scale) that represent the two other fields (refer to Fig. 1.20).

Now assume the charge at C is $+5 \times 10^{-6}$ coulombs. Find the net vector field at point B that the charge at B would be acting in. (To find this field at point B, merely remove the charge at B.)

$$\boldsymbol{E}_{BA} = (Q_A/r^2)\,(9 \times 10^9) = 0.90 \times 10^4\; a_{rBA}\; \text{V/m}$$
$$\boldsymbol{E}_{BC} = (Q_C/r^2a)\,(9 \times 10^9) = 2.25 \times 10^4\; a_{rBC}\; \text{V/m}$$

Again, solve graphically, for the solution, as shown in Figure 1.21.

The actual solution, using vector notation, is found to be $\boldsymbol{E}$ (at B) equals $1.73 \times 10^4 \angle 66.5°$ volts/meter.

Line Capacitance

Another interesting problem from field theory is that of finding the capacitance between two parallel wires due to the field of each wire. For a single long wire conductor, the electric field about it is radial and is given by (outside the conductor), $E = Q/(2\pi, r)$, where r is the radial distance away from the center of the conductor. The potential at some point, b, away from the center (farther than the radius, a, of the conductor itself), is given by

$$V = \int_a^b \text{E} dr = \int_a^b [Q/(2\pi\varepsilon r)]dr = [Q/(2\pi\varepsilon)]\ln(b/a). \qquad (\mathbf{1.20})$$

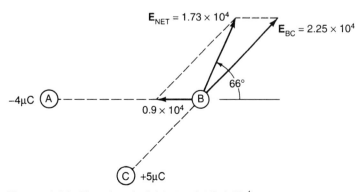

Figure 1.21 The electric field at point B (×10⁻⁴)

On the other hand, for two parallel wires, charges on the other wire contribute equally (for the same amount of charge but of opposite sign), and the total potential difference is twice that produced by the charge on one wire. Also, by definition, $C = Q/V$, so the capacitance, C (with C', or c, being C per unit length) becomes

$$C' = (\pi\varepsilon)/[\ln(b/a)] \text{ farads/meter.} \tag{1.21}$$

This relationship is important in transmission line theory and will be discussed where appropriate.

Magnetic Quantities and Definitions

Magnetic effects are related to the motion of charges, or currents. Electric current, on the other hand, can be thought of as moving charges. The definition of current is: the time rate of change at which the electric charge passes through a surface area per unit of time. As concisely stated by Smith (*Circuits, Devices and Systems*, 4th ed., Wiley, 1983), "The current is the net rate of flow of positive charges, a scalar quantity. In the specific case of positive charges moving to the right and negative charges to the left, the net effect of both actions is a positive charge moving to the right; the instantaneous current to the right is given by

$$i = dq^+/dt + dq^-/dt$$

In a neon light, for example, positive ions moving to the right and negative electrons moving to the left contribute to the current flowing to the right."

From the concept of moving charges, one can begin to understand magnetic fields. For permanent magnets (due to "static" magnetic fields), recall from physics that for some materials, the molecular structure has the electron orbits of the atoms aligned. These tiny moving charges of electrons produce tiny currents. This alignment results in magnetic fields; actually, ferromagnetic materials can be thought of as a large number of magnetic domains, with the domains being mostly aligned. However, when these (magnetic) domains are in disarray or randomly aligned, the material will be unmagnetized.

Oersted, in 1819, observed that a magnetic flux existed about a wire carrying an electric current. A few years after Oersted observed the effect of magnetic flux, Ampere found that wire coils carrying a current acted in the same manner as magnets. Or, simply stated, a coil of several turns of wire produced a stronger magnetic flux than one turn of wire in the coil for the same current. And, if there were a ferromagnetic material to carry (or to provide a path for) the magnetic flux, the flux strength would be much greater.

Consider a toroidal ferromagnetic ring with a coil of several turns of wire wrapped around it and the coil connected to a variable current source. Assume the current is zero and the material is not magnetized to begin with. Then for an increase of current, a nonlinear increase in flux results. As the current is further increased, the flux tends to level off. The leveling off is due to most of the magnetic domains in the material having aligned themselves in the same direction; consequently, increasing the current beyond the "knee of the bend" does not significantly increase the flux. If the current is now decreased toward zero, most (but not all) of the magnetic domains return to random directions for zero current. The "going up" path is not necessarily the same as the "going down" path; these paths are known as the hysteresis curve for a particular material. On the "coming down" side, when the current reaches zero the material is left partially magnetized, and

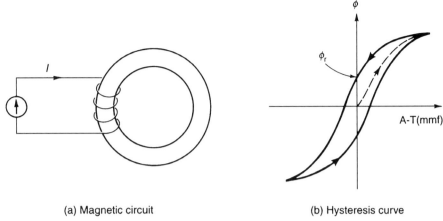

(a) Magnetic circuit   (b) Hysteresis curve

Figure 1.22 Magnetic characteristics

this is called the residual magnetism. This product of the current and the number of turns of wire, or amp-turns (A-t), is frequently called the "magneto-motive force," (mmf), or sometimes indicated with a script F. The mmf is the source that causes the magnetic flux (Φ). A plot of these two quantities is given in Figure 1.22b; this plot is referred to as the hysteresis curve. If the "fatness" of the hysteresis curve were small (*i.e.,* the residual magnetic flux is small), only the first quadrant of the hysteresis need be used and drawn as a single line; then, only one curve per kind of magnetic material is needed.

Before a numerical problem can actually be solved, you need to quantify and further define several terms. Rather than using flux, Φ (units being webers), it is more appropriate to use flux density, B (with units being tesla or webers/unit area). Also, rather than using the straight magneto-motive force designation, a more useful one is the mmf/length, which then becomes the magnetic field intensity, H. These "standardized" quantities will allow for a plot of B vs H; this plot is always given for solid magnetic materials (*i.e.,* before any air gap might be "cut" into them). Figure 1.23 shows these two typical curves (along with a straight line plot for air, or free space, indicated by μ_o).

It is also interesting to note that the magnetic flux in an air gap (in series with the magnetic circuit) is concentrated and may be very high. The lines of magnetic flux produced in the material and the air gap are the same and continuous when there is a current in the wire. Without the ferromagnetic material in the path, the relationship between the current (causing the magnetic flux) and the resulting

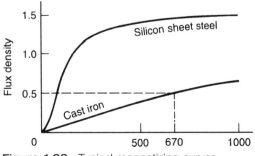

Figure 1.23 Typical magnetizing curves

magnetic field would be linear but very weak. The linear constant of proportionality is called the permeability of free space, and is given as μ_o.

$$\mu_o = 4\pi \times 10^{-7} \text{ H/m} \qquad (1.22)$$

For ferromagnetic materials, the relative permeability is nonlinear; for approximate calculations, it is sometimes linearized in the region of the curve before magnetic saturation is reached. It is interesting to note that the slope of the curve in the saturation region approaches that of free space or air, thus

$$\mu_r \mu_o \rightarrow \mu_o.$$

The relationship between B and H may then be expressed as the slope of the B vs H curve,

$$\mu = \mu_r \mu_o = B/H. \qquad (1.23)$$

As stated previously, flux lines are continuous (*i.e.*, the lines of flux in the ring and air gap are the same), whereas the flux density in the ring may or may not be the same as that in the air gap (frequently they are considered the same by neglecting the fringing effect in the gap). Also, the cause of the flux (*i.e.*, the amp-turns) is thought of as being distributed along the whole ring, thus it is appropriate to use field intensity, H (amp-turns/meter or mmf/m), A-t/m; the length, for this example, is merely the mean circumference of the ring (the width of the air gap usually being negligible). The net mmf in a magnetic series loop equals zero; stated another way, the mmf source must equal mmf drops (or losses) in the series circuit. Thus, summing these mmf drops in the iron and air gap is all that is needed is to find mmf for the source.

A numerical example follows: Assume a coil of 50 turns of wire wrapped around a toroidal cast iron ring whose ring radius (to the center line of the ring) is 4 cm and has a cut (air gap) in it of 1 mm; the cast iron ring itself has a radius of 2 mm (see Fig. 1.24). What current is necessary to produce a flux density in the air gap of 0.5 tesla?

Finding the mmf necessary to produce a flux density in the air gap involves applying Equation 1.22 for a required flux density in the air gap of 0.5 T,

$$H_{AG} = 0.5/(4\pi \times 10^{-7}) = 0.398 \times 10^{+6} \text{ A-t/m}.$$

Because the $H = $ mmf/l, the mmf$_{AG} = H \times$ length of air gap,

$$\text{mmf}_{AG} = (0.398 \times 10^6)\,(1 \times 10^{-3}) = 398 \text{ A-t}.$$

(Here, since the area of the air gap is the same as the cast iron area face plates, it won't be necessary to find the flux. It should be noted that if fringing were to

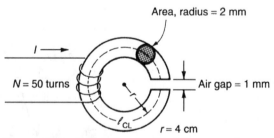

Figure 1.24   Magnetic circuit for the numeric example

be considered (where the air gap "bulging" would cause the area to be increased), the flux would need to be found. This is so because flux in a series circuit would be the same and continuous; and the flux densities of each portion of the path would be different.) Finding the necessary mmf needed to produce the same flux density in the cast iron is merely that of taking the data from the graph in Figure 1.23; for a B of 0.5 T, the graph yields an H of 670 A-t/m. The mean length of the cast iron path is $2\pi r = 2\pi(4 \times 10^{-2}) = 0.251$ m. Thus, mmf needed for the cast iron is $mmf_{CI} = (H_{CI})(\text{path length}) = (670)(0.251) = 168$ A-t, the mmf needed by the source is

$$mmf_{coil} = \Sigma mmf's = mmf_{AG} + mmf_{CI} = 398 + 168 = 566 \text{ A-t.}$$

Therefore the coil current $i = mmf/\#turns = 566/50 = 11.3$ A.

Note that while the air gap is only a fraction of the length of path for the flux, most of the mmf is used for this portion of the magnetic circuit!

Another method of solution uses a "linearized curve" expressed as a relative permeability. This method will be demonstrated and involves the concept of summing the various reluctances in a series magnetic circuit. (Here, you can think of summing the reluctance, $\Re$, in a path like summing resistance in an electrical series circuit). Similar to Ohm's law in an electrical series circuit where $I = V/(\Sigma\Re's)$ or $V = I/(\Sigma\Re's)$, the flux in a magnetic series circuit is $\Phi = mmf/(\Sigma\Re's)$ or $mmf = \Phi(\Sigma\Re,)$. For most magnetic circuit problems using this method of solution, the relative permeability is given; however, to compare methods, this value will be calculated from the graph (Fig. 1.23). Since μ is B/H,

$$\mu = \mu_r\mu_o = B/H = (0.5)/670 = 7.46 \times 10^{-4} \text{ H/m}$$
$$\mu_r = \mu/\mu_o, \ \mu_r = (7.46 \times 10^{-4})/(4\pi \times 10^{-7}) = 5.94 \times 10^2.$$

Again, like finding the physical resistance of a wire in an electric circuit, you may find the reluctance, $\Re$, of a magnetic circuit with the same type of formula.

$$\Re = (1/\mu)(\text{length of path/cross-sectional area})$$
$$\Re_{CI} = [1/(7.46 \times 10^{-4})][(8\pi \times 10^{-2})/(4\pi \times 10^{-6})] = 26.8 \times 10^6 \text{ A-t/wb}$$
$$\Re_{AG} = [1/(4\pi \times 10^{-7})][(1 \times 10^{-3})/(4\pi \times 10^{-6})] = 63.4 \times 10^6 \text{ A-t/wb}$$
$$\Sigma\Re's = (26.8 + 63.4) \times 10^6 = 90.2 \times 10^6 \text{ A-t/wb.}$$

The flux is the same for all parts for the series circuit and may easily be found from the flux density required in the air gap.

$$\Phi = BA = 0.5(4\pi \times 10^{-6}) = 6.28 \times 10^{-6} \text{ webers.}$$

The mmf needed is then

$$\Phi = mmf/\Sigma\Re's, \quad mmf = \Phi(\Sigma\Re's) = (6.28 \times 10^{-6})(90.2 \times 10^6) = 566 \text{ A-t}$$

and the current needed in the coil is

$$i = (mmf)/(\#turns) = 566/50 = 11.3 \text{ A.}$$

Although the previous example involved a coil of wire, a more fundamental problem would be that of a straight long wire. The flux density at some radial distance, r, from the center and perpendicular to the line of the wire is given as

$$B = (\mu i)/(2\pi r) \qquad \textbf{(1.24a)}$$

Now, if the long wire were formed into a circular coil with a radius r, you could use calculus and consider a differential length of the wire and integrate around the closed loop to find the flux density within the loop. If, on the other hand, there were several turns, N, for the loop, the equation for the flux density would be

$$B = (\mu Ni)/r \qquad (1.24b)$$

Line Inductance

An interesting derivation for the inductance of a transmission line (which will be addressed in a later chapter) will follow. The inductance is a measure of the reactive voltage drop along a line, and it may be found by considering the magnetic field(s) about a current-carrying wire(s).

First consider a single long wire (that is, the return conductor is assumed to be a long distance away). The strength of the magnetic field around the wire is given by the integral of a closed path around the current-carrying conductor,

$$\int H \,\text{dot}\, ds = i \qquad (1.25a)$$

(where s is the distance along the path of integration). For H having a tangential direction to the assumed symmetrical rings around the wire and the integral of ds as the circumference of a circle with radius, r, the current i will be related to the scalar equation as

$$\int H \, ds = i = H \int ds = 2\pi r H. \qquad (1.25b)$$

Recall the flux density, B, is related to H by the permeability, μ, then B is given by

$$B = (\mu i)/(2\pi r).$$

The inductance, L (with L', or l, being L per unit length), is related to flux linkages by $L = (N\Phi)/i$ (here again, N = 1 for a long straight wire), and the flux per unit length is given by

$$\Phi = \int_a^b [(\mu i)/2\pi r]dr = [(\mu i)/2\pi r]\ln(b/a) \qquad (1.26)$$

where a = radius of the conductor and b = distance away from the centerline of the conductor. This equation neglects the partial flux linkages within the conductor, which, if included, would add an additional term of $\mu i/(8\pi)$. The inductance may then be found for several configurations.

The simplest relationship would be for a *coaxial cable* configuration (here the outer conducting tubular surface is located at radius b),

$$L' = [(\mu)/(2\pi)]\ln(b/a) + \mu/(8\pi) \text{ henry/meter.} \qquad (1.27)$$

For a two-conductor line, both lines produce a magnetic alternating flux around each line and each line has induced into it (by the other line) a reactive voltage as the flux linkages change. For this two-wire line, the inductance per unit length may be shown to be

$$L' = (\mu/\pi)\ln(b/a) + \mu/(4\pi) \text{ henrys/meter.} \tag{1.28a}$$

Assuming a relative high frequency (inner and outer conductors flux densities become small because of skin effect) and with $\mu_o = 4\pi \times 10^{-7}$, gives

$$L' = 4 \times 10^{-7}\ln (b/a) \text{ henrys/meter.} \tag{1.28b}$$

In general, as the separation distance between two wires increases, the inductance increases while the capacitance decreases.

For many of the concepts presented in this section, the use of vector notation was minimized; however in an actual problem, you could tailor and minimize the solution steps by your own ability using this notation. For more comprehensive vector notation relationships, please refer to some of the problems elsewhere in this book.

RECOMMENDED REFERENCES

Bobrow, *Elementary Linear Circuits Analysis*, 2nd Edition, Holt, Rinehart and Winston, 1987.

Jones, Lincoln and Howard Smolleck, *Electrical Engineering: FE Exam Preparation*, 3rd Edition, Kaplan AEC, 2005.

Nilsson, *Electric Circuits*, Prentice Hall, any edition.

Machines

This chapter on electrical machines covers three main areas of study: transformers, dc machines, and ac machines. Each area covers several subtopics. For example, under ac machines, synchronous and induction machines and an introduction to ac control motors are presented.

Starting the review with transformer theory may seem unimportant, but many ac rotating machine circuits are based on the non-ideal transformer equivalent circuit. In the author's opinion, if you have trouble understanding and using the basic non-ideal transformer equivalent circuit, you should probably not spend time reviewing the section on rotating ac machines, but instead use your study time on other subjects. However, if you have little familiarity with equivalent circuits, it would pay to use a little of your study time to work through this transformer area.

While the depth of coverage in this chapter is necessarily limited, the study problems in the companion book, *Electrical Engineering: Problems and Solutions*, provide additional depth of review and practice.

TRANSFORMERS

This discussion will deal with transformers used in ac circuits with sinusoidal waveforms (in contrast to those used in pulse and other specialized types of circuits). The analysis will start with the ideal transformer (introduced in Chapter 1), which is considered to be lossless with 100 percent efficiency. For these ideal devices, the product of the input volt-amps is presumed to equal the product of the output volt-amps. The product, *VA*, is the apparent power (*VA*) rather than real power (watts). The larger the volt-amp product, the larger the transformer (for the same frequency). The name plate rating of the transformer gives the normal operating conditions;

these ratings include the frequency, the voltage and the voltage ratio, and the *VA* (or *kVA* or *MVA*) rating. The voltage ratios are the same as the turns ratio, "*a*", and the current ratios are inversely related to the turns ratio. For this discussion, assume the left side is primary (1) and the right side is secondary (2). For example, assume a transformer with a name plate rating of 5 kVA, 60 Hz, and 880/220V (see Fig. 2.1). The primary side might come from an 880 volt source and the secondary would be at 220 volts. The voltage/current rating is always given in rms values, and the turns ratio, of course, would be 4:1. The current (or load) on the secondary side could be as high as $VA/V_2 = 5,000/220 = 22.7$ amperes, while the primary side would be $I_1 = 5,000/880 = 5.68$ amperes, or 1/4 of 22.7 amperes. It is interesting to note that a resistive load on the secondary side is $R_L = V/I = 220/22.7 = 9.69$ ohms, while it would be $880/5.68 = 155$ ohms on the primary side. This equivalent resistance reflected on the left side is $(a)^2 R_L$.

$$R_{L\text{-eq}} = (a)^2 R_L = (4)^2 \times 9.69 = 155 \ \Omega.$$

Here, the $kVA_{in} = kVA_{out}$ equals $kW_{in} = kW_{out}$ since the load is a purely resistive one. On the other hand, if the load were $5 + j5$ (or $Z_L = \sqrt{2} \times 5\angle 45°$), the current would be

$$I_2 = V_2/Z_L = 220/(\sqrt{2} \times 5/45°) = 31.1\angle -45° \text{ A,}$$

and the current on the left side is $I_1 = I_2/a = 7.78$ amperes. These exceed the name plate values, and a larger transformer would normally have to be selected. The equivalent impedance of the load if reflected to the left side is given as

$$Z_{L\text{-eq}} = (a)^2 Z_L = (4)^2 5 + j(4)^2 = 80 + j80 \ \Omega$$

Caution is needed here because the power-in/power-out relationship is misleading, $P = VI\cos \Theta = 220 \times 31.1 \times 0.707 = 4837$ W, or less than 5 kW; again, the transformer rating is *kVA*, not necessarily *kW*!

For a non-ideal transformer, efficiency ratings are, of course, less than 100 percent and are found much like those of motors, generators, and alternators, except for no friction and windage loss. A short review of the non-ideal transformer relationships, including actual voltages and currents, follows. The power losses include the real windings (copper loss) and the losses due to the hysteresis and eddy current effect (iron loss). Both the primary winding and secondary windings that are represented by copper losses also have inductances that must be included in a model of an actual transformer and be represented as such. However, since the current and voltages levels are usually different for each side of the transformer, one or the other side may be referred to by the a^2 or $1/a^2$ term, thus enabling the

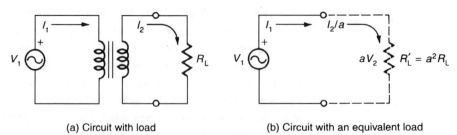

(a) Circuit with load   (b) Circuit with an equivalent load

Figure 2.1 A typical two winding loaded transformer

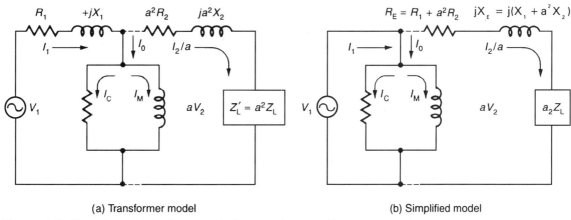

(a) Transformer model   (b) Simplified model

Figure 2.2 Equivalent circuit of a two-winding transformer with respect to primary

model to be shown with one circuit. As an example, assume everything, including the load, is referred to the primary side as indicated in Figure 2.2.

The resistances of the coil wires are indicated by $R_{eq} = R_1 + a^2 R_2$ and the same form is true for the inductances of each side. Also, the model should include a path to account for iron loss (or core loss), I_c, and a magnetizing current (energy storage), I_M. These two currents are usually combined into a parallel branch at the center of the model; and the phasor sum is called the exciting current, I_o. The exciting current is small compared to the rated current, therefore this parallel branch may be moved to the extreme left of the model with very little error. This allows one to combine the series components for a much simplified circuit model (see Fig. 2.2 b). The values for this model are easily obtained from an actual transformer by a series of open and short circuit tests. The net result of having this transformer model is that one can easily compute voltage regulation and efficiency for most any condition.

As an example, to find the efficiency of the transformer for any particular load, one can start from either end of the simplified model and assume, for that end, the voltage is known. Say the secondary, or output voltage is at rated value, then the approximate input voltage, of course, will be given by the turns ratio, but the exact value needs to be computed, based on what the load current happens to be. The computed input (see Fig. 2.3) voltage must be the referred value (to the primary side) of the output voltage $(V_2' = a V_2)$ plus all of the series voltage drops in the path (here, the current is the referred value of $I_2' = I_2/a$; it is determined by the load impedance) as follows:

$$V_1 = V_2' + (R_E + jX_E)I_2' \qquad (2.1)$$

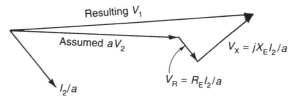

Figure 2.3 Transformer phasor diagram

Knowing the input voltage, the power losses in the series resistors and the parallel branch are easily computed. However, the computed voltage V_1 allows one to also compute the voltage regulation (usually under full load conditions) easily.

$$\%\text{V.R.} = 100(V_{1@\text{no load}} - V'_{2@\text{full load}})/(V'_{2@\text{full load}}) \tag{2.2}$$

Transformer per Unit Values

Frequently, it is advantageous for large transformers, especially if several transformers are involved, to use per unit (pu) or normalized values for the transformer parameters. These normalized values are nothing more than the percent of the various types of variables as related to some base quantity; these could be base voltage, base current, base impedance, and base power.

As an example, if a transformer is rated at 100/400 kV but has an actual primary voltage of 102.5 kV then its pu value is given as 102.5/100 = 102.5% or 1.025 pu. For the secondary side, if the actual voltage is 395 volts, its pu value is 0.9875. If the kVA rating is 250 kVA and the load is, say, 150 kW along with a $\text{kVA}_{\text{reactive}}$ of 100, then, combining powers, $\sqrt{150^2 + 100^2} = 180.3$ kVA. The loading will then be 180.3/250 = 0.721 pu; or, in terms of complex power, **S**, it is **S** = (150 + j100)/250 = 0.6 + j0.4 pu. By reducing the various parameters to their pu values, it makes the "bookkeeping" easier to manage and gives a better perspective of the utilization of the transformer(s), or any other device in a system.

DIRECT CURRENT MACHINES

The first portion of this presentation is very elementary and is written for those whose undergraduate days did not include any machinery; these first few pages should only be scanned by those with a background in this subject. It is important, however, to at least review the basic concepts of this area, as questions in control system are frequently based on controlling some kind of dc machine (including linear motors and disk drives).

As previously pointed out (in Chapter 1), a conductor moving in a magnetic field will have a voltage induced in it. The voltage will be greatest when normal to the magnetic flux and least as it approaches the direction of the flux. The magnitude of the voltage is proportional to the length of the conductor within the field, the velocity of the conductors, and the strength of the magnetic field. Recall that the magnetic field may be due to a permanent magnet or may be produced and controlled by current in a wire, usually a coil of wire wrapped around a ferromagnetic material (here called "field poles"). The flux density in the ferromagnetic material is related to the current, I_f, by the hysteresis curve and the air gap. For example, a pair of conductors rotating in a magnetic flux (normal to each other) will produce a voltage, E_g, caused by a field current, I_f, (see Fig. 2.4). Any current that might flow because of the generated voltage gets to the outside circuit through commutator segments and brushes (as in Fig. 2.4c). The magnitude of the voltage generated is then a function of field current, I_f, and the rotational velocity, because all other parameters are fixed. An actual dc generator, of course, has many more conductors and commutator segments, and frequently more pairs of poles. The connections from the armature conductors to commutator segments and the number of pairs of poles are important considerations in terms of the voltage/current ratings; however, for our purpose, these details will not be considered. If the field current is supplied by a separate source, a typical plot of the generated

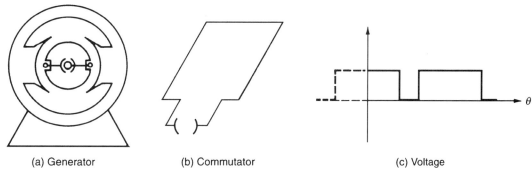

(a) Generator (b) Commutator (c) Voltage

Figure 2.4 Simplified two pole dc generator

voltage is given in Figure 2.5. Note that the voltage curve starts leveling off as the field current increases; this, of course, is due to the magnetic saturation of the ferromagnetic material, including the field poles, the armature, and the frame, or yoke, of the machine. Generally the armature resistance, including the brushes (actually the brush resistance is nonlinear), is a relatively low resistance, perhaps a fraction of an ohm, whereas the shunt field is relatively high, perhaps 25 to 100 ohms. If the armature current is small, the $I_a R_a$ drop is negligible. The implication is that the terminal voltage is very near the generated voltage, E_g, and is referred to as a no load voltage, V_{NL}. Conversely, the reason for the existence of a generator is to provide relatively large currents to a load, where V_T now becomes the full load voltage, V_{FL}. The concept leads to the definition of voltage regulation. The percent voltage regulation is defined as

$$\%VR = [(V_{NL} - V_{FL})/V_{FL}] \times 100 \tag{2.3a}$$

$$\%VR = [(E_g - V_{Fl})/V_{Fl})] \times 100. \tag{2.3b}$$

Consider the following example: Assume a dc generator is being driven by a prime mover at full-rated speed. The rated output voltage is to be 200 volts when the rated armature current of 50 amperes flows to the load. If armature resistance is known to be 0.2 ohms, what field current, I_f, is necessary and what is the voltage regulation from no load to full load?

Knowing that, in this particular case, the no load voltage, E_g, minus the voltage drop in the armature resistance must be the full load terminal voltage, V_T, the no

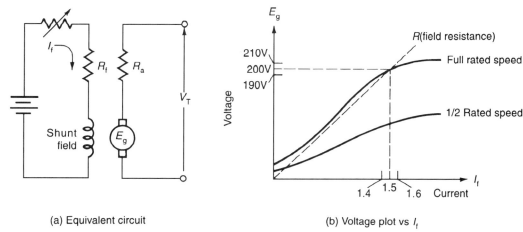

(a) Equivalent circuit (b) Voltage plot vs I_f

Figure 2.5 Typical generated voltage curve vs field current

load voltage is easily calculated to be 200 + (50 × 0.2) = 210 volts. Thus from Equation 2.3b, the voltage regulation is

$$\%\text{VR} = [(210 - 200)/200] \times 100 = 5\%.$$

In other words, the measured voltage across the load, R_L, is 200 volts, but if the load resistance is removed, the terminal voltage would jump to 210 volts. From Figure 2.5b, the field current required is 1.6 amperes.

Recall that when there is a current flow in the presence of a magnetic field, a force is exerted on the conductor(s), and the torque on the armature conductors opposes the direction of the shaft rotation. As the current from the generator increases, the counter torque increases, making the prime mover work harder. If the electrical power dissipated in the external load resistor is the power output, and the mechanical horsepower (from the prime mover) to the shaft is the power input, then one may easily determine the efficiency of the machine. The calculation of efficiency has some interesting aspects. In the previous example, the electrical power delivered to the load was $V_T I_a = 200 \times 50 = 10$ kW. If it is known that the power taken from the prime mover is 16 hp at rated speed, the efficiency may be found. After converting all powers to watts (16 hp × 746 watts/hp = 11,936 watts), the efficiency calculation is

$$\text{Efficiency} = P_o/P_{in} = 10,000/11,936 = 0.838 = 83.8\%$$
$$= P_o/(P_o + \text{all losses within}).$$

Once able to identify the losses within the machine (here, almost 2 kW), one may begin to find a reasonable model for the dc generator. Two electrical losses are the power lost to provide for a magnetic field, which is $I_f^2 R_f$, and the power lost in the armature windings, which is $I_a^2 R_a$; these two losses are sometimes lumped together and called the "copper loss" (see Fig. 2.6). Still another loss may be found by driving the prime mover at rated speed with all electrical connections removed; the measured power taken from the prime mover by the electrical machine is called the "friction and windage" loss, F&W. There is still another loss that is difficult to measure. This loss is sometimes called the "iron loss" and is mostly due to both a hysteresis and eddy current effect caused by the armature rotating in the presence of a magnetic field. Efficiency may be now defined in a slightly different manner (see Eq. 2.4).

$$\%\text{Efficiency} = (P_o/P_{in}) \times 100 \tag{2.4a}$$

$$\%\text{Efficiency} = P_o/(P_o + \text{Cu} + \text{Fe} + \text{F\&W}) \times 100 \tag{2.4b}$$

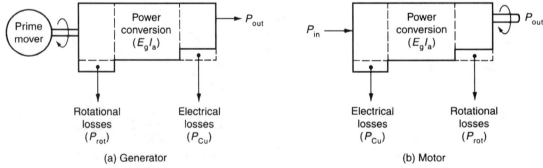

(a) Generator (b) Motor

Figure 2.6 Power losses for a dc machine

For example, from the previous problem, determine the iron loss if it is known that friction and windage loss was measured to be one horsepower at rated conditions. The copper loss is easily calculated to be $I_f^2 R_f + I_a^2 R_a$, or $1.6^2 \times 40 + 50^2 \times 0.2 = 602$ watts. The F&W loss is 746 watts; therefore, the difference between these losses and all losses must be the iron loss $1,936 - (746 + 602) = 588$ watts. In practice, for small speed changes, the iron loss and the friction and windage losses are usually lumped together and called the rotational losses.

The equivalent circuit of a dc machine may take several forms such as the ones shown in Figures 2.5a, 2.7a, and 2.8a. With these circuits one is usually dealing with speeds in rpm and horsepowers in or out. However, when working with torques, the preferred equivalent circuit frequently is changed to a dependent type circuit model such as those shown in Figure 2.7a, b.

Here, because of the simplicity of power (in watts) being the product of torque (in N-m) and speed (in rad/s), the new model combines a mechanical schematic with the electrical portion. For the machine parameter, $K\Phi$ or $K'\Phi$, are found exactly the same way as in the electrical model except the speed is given in rpm for K and in rad/sec for K' ($K' = K60/2\pi$).

The steady state relationships are given as

$$E_g = K'\Phi \quad \text{and} \quad T_d = K'\Phi I_a$$

where

E_g = the back or generated emf
Φ = the air gap field flux (constant for a fixed field)
ω = speed in radians/second
T_d = developed torque in Newton-meters
I_a = armature current (amperes)
K or K' = a constant for a particular machine
D = mechanical damping (representing rotational losses)

As an example, consider a fixed field dc motor as in Figure 2.5b. Assume the rated speed is 100 rad/s, the rated voltage is 100 volts, the rated current is 50 amperes, and the rated output power is 5 hp. It is known that the armature resistance is 0.1 ohms. What is the equivalent amount of torque being consumed by the rotational losses?

$$P_o = (5 \text{ hp})(746 \text{ w/hp}) = 3,730 \text{ watts}, \quad T_o = P_o/\omega = 3,730/100 = 37.3 \text{ N-m}$$
$$Eg = V_T - I_a R_a = 100 - 50 \times 0.1 = 95 \text{ volts}$$
$$K'\Phi = E_g = 95 = K'\Phi \, 100, \quad K'\Phi = 95/100 = 0.95$$
$$T_d = K'\Phi I_a = 0.95 \times 50 = 47.5 \text{ N-m}$$
$$T_D = T_{\text{dev}} - T_o = 47.5 - 37.3 = 10.2 \text{ N-m}.$$

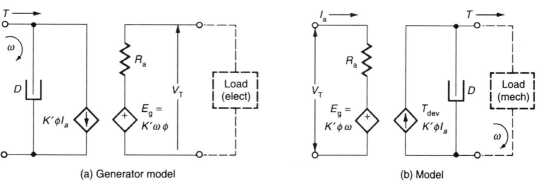

(a) Generator model   (b) Model

Figure 2.7 A dc machine model involving torques

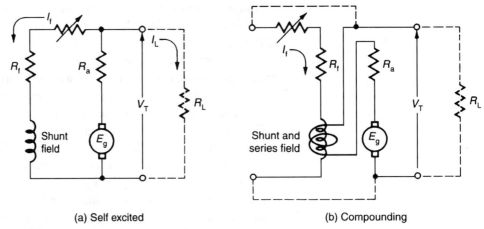

(a) Self excited   (b) Compounding

Figure 2.8 Various connections for a dc machine

As a check, the reader may wish to subtract the sum of all power losses from the power input to see if the results will check with the known power output.

It is important to understand several modifications to the circuit of the basic dc machine. One such modification is to provide for its own shunt field current. Here, the developed terminal voltage is used to furnish this current such that a separate voltage source for the field is not needed. The modified circuit configuration is called a "self excited" shunt field connection. This may be accomplished by taking advantage of the residual magnetism in the shunt field poles, which produces a small voltage in the armature (see Fig. 2.5b). This assumes the circuit connections are correct, in that I_f will flow in a direction to produce more generated voltage, rather than opposing the residual voltage. This voltage will increase until E_g intersects the slope of the dashed resistance line shown in Figure 2.5b. Another modification, which improves voltage regulation, is to increase the magnetic field proportional to the load current.

This operation is called "compounding" and is achieved by adding a few turns of wire to the shunt field (see Fig 2.8b); these added turns are called "series field windings." When the series field aids the shunt field, the configuration is called "cumulative compounding;" if the series field opposes the shunt field, it is called "differentially compounding." The dc machine used as a motor involves very little change in theory. The only difference between a generator and a motor is that one puts in electrical power and gets out mechanical shaft power. The power output equals the electrical power input less all losses within the machine (see Fig. 2.9b).

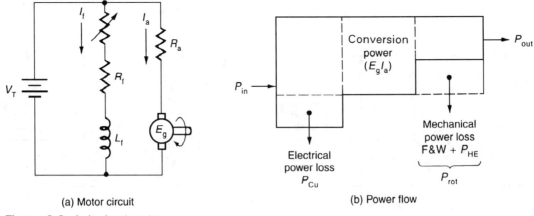

(a) Motor circuit   (b) Power flow

Figure 2.9 A dc shunt motor

The torque developed, T_d, within the machine is a function of the magnetic flux and the current in the armature; it can also be shown that it is related to the product of $I_a E_g$. It should be noted that this torque is not a direct function of speed. As the torque causes the shaft to turn, a back emf (*i.e.*, again, the generated voltage, E_g) is also developed; this emf is a function of speed, n, and the magnetic field strength. Consider a motor with a fixed shunt field. Ignoring the field temporarily, the loop around the armature portion of the circuit is given by

$$V_T = I_a R_a + E_g = I_a R_a + K_g n. \tag{2.5a}$$

Solving for speed, n, yields Equation 2.5b,

$$n = (V_T - I_a R_a)/K_g. \tag{2.5b}$$

From Equation 2.5b it is seen that for a constant supply voltage and fairly large changes of torque requirements (proportional to I_a), the speed change may be very small. This is because V_T is usually large compared to the $I_a R_a$ drop. However, the "constant" K_g is only constant for a fixed field; the magnetic flux is easily changed with a change of field current, I_f. Thus a full equation reflecting a variable field current is

$$n = (V_T - I_a R_a)/K I_f. \tag{2.5c}$$

Equation 2.5c indicates that the speed is easily controllable over a fairly large range with changes of field current, I_f. There is a limit to the practical maximum value of I_f because of saturation. It may be seen from Equation 2.5c that, if upon starting, $n = 0$ (causing $E_g = 0$), then $V_T = I_a R_a$; when the speed is zero, I_a could become extremely high. In fact, for larger machines, one usually uses some kind of controller to limit the starting current. The controller may be nothing more than temporary insertion of an external starting resistance in series with the armature, or it could be some kind of resistive solid state device.

To determine the efficiency of the motor (see Fig. 2.6b), one needs to know the power input, which is easily found from the power output plus all losses. For the running motor, the generated voltage, E_g, is less than the terminal voltage by an amount $I_a R_a$, $E_g = 200 - (50 \times 0.2) = 190$ volts, which yields a field current (from Fig. 2.5b) of 1.4 amperes. The $I^2 R$ losses are $(1.4)^2 \times 40 = 78.4$ W plus $(50)^2 \times 0.2 = 250$ W, or $P_{Cu} = 328$ W; and the F&W loss was 746 W. The other loss was the iron loss (hysteresis and eddy current effect), which was 588 W. The efficiency may now be determined as

$$\%\text{Eff} = P_o/P_{in} = [7{,}460/(7{,}460 + 746 + 328 + 588)] \times 100 = 82\%.$$

AC ROTATING MACHINES

This section on ac rotating machines will start with an elementary overview of the kinds of machines that could possibly be included in a particular examination. Following the overview, a number of detailed problems will be presented that will involve a wider scope of material beyond this short review.

For the following discussion involving three-phase machinery, almost all of the theory is based on per-phase relationships (using one leg of a wye to neutral) and should be readily understandable. However, if the reader is on "shaky" ground, it is recommended he or she first read the section on Three-Phase Review, presented at the beginning of Chapter 3.

Of the many kinds of ac machines, the most common are (1) the synchronous machines (as a motor or as an alternator-generator), (2) the induction motors, (3) the series motors, and (4) specialized machines (*i.e.,* stepping motors, servo motors, etc.). Most ac machines involve a rotating magnetic vector that rotates at a constant speed and is directly related to the line frequency and inversely related to the number of pairs of poles, #PP (the stator windings are spread out to emulate a certain number of pairs of poles). As an example, if the stator winding is excited from a 60 Hz source, and the winding pattern emulates a single pair of poles, the rotating magnetic vector rotates at exactly 3,600 rpm. If the frequency were doubled, the speed would be 7,200 rpm. Alternatively, if the windings were arranged to emulate two pairs of poles, the speed would be half as much. If the machine is three-phase, the direction of the rotating magnetic vector depends on the sequence of the phase voltages. To reverse the direction, merely interchange any two of the three lines going to the machine.

Synchronous Machines

The synchronous machine is usually a three-phase device with the stator windings creating the rotating flux vector (see Fig. 2.10). Think of this vector as if it were a magnet rotating inside the machine; if one were to place a rotor which was also a magnet in the presence of the rotating vector, it would "lock in" with the rotating flux vector. The imaginary rotating magnet with N-S poles and the rotor with S-N poles would attract each other and both would rotate at exactly the same speed, or in synchronism. Normally, the rotor magnetic has a field strength that is controllable from an adjustable dc source, much like the shunt field. One should realize that the synchronous machine operates at an exact speed, which usually requires an additional dc source to control its field. The synchronous machine may be used as a motor (here, the power factor may be controlled by the dc field current). The machine may also be used as an alternator, to generate a three-phase voltage whose voltage output may also be controlled. Of course there are many facets to the theory of operation, such as how to get the machine up to synchronous speed, or how to get a leading or lagging power factor. The equivalent circuit of a synchronous generator in its simplest form on a per-phase basis is given in Figure 2.11.

Note that an impedance X_s is introduced; also note that an induced (or generated) voltage, E_f, appears as the source. The E_f is a function both of synchronous speed and the dc field current of the rotor; and Z_s is a "fictitious" quantity, called

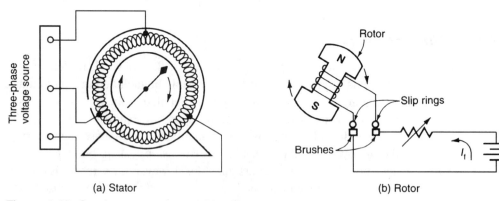

(a) Stator   (b) Rotor

Figure 2.10 Synchronous motor configuration

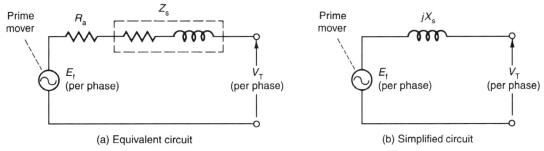

Figure 2.11 Equivalent circuit for a synchronous generator

the synchronous impedance. This synchronous impedance is due, in part, to the magnetizing and leakage flux caused by both the field and armature currents. The resistive portion is considerably smaller than the reactance, and, without much loss of accuracy, $Z_s + R_a = X_s$, or synchronous reactance.

As an example, assume the line (rated) voltage is 15 kV, the synchronous reactance is 10 Ω/phase and a load is $20 + j20$ Ω/phase (assume a wye configuration). Compute the line currents and the power output. From the circuit of Figure 2.12a, the phasor diagram is obviously that of Figure 2.12b; here the line current is $\mathbf{V}$/phase/$(20 + j20) = (15/3\angle 0°)/ (20/2\angle 45°) = 0.307\angle{-45°}$ kA.

$$E_f/\text{phase} = jX_s\mathbf{I} + \mathbf{V}/\text{phase} = (j10)(0.306 \times 10^3\angle{-45°})$$
$$+ 8.67 \times 10^3\angle 0° = 11.03\angle 11.3° \text{ kV}$$
$$P_o/\text{phase} = VI\cos\theta = (8.67 \times 10^3)(0.306 \times 10^3)\cos 45° = 1.88\text{Mw}$$
$$P_o(\text{Total}) = 3P_o/\text{phase} = 3 \times 1.88 = 5.64 \text{ Mw.}$$

The angle, θ, is the power factor angle and zeta, ζ, is the power (torque) angle. For the synchronous motor operation, the theory is essentially the same as for the generator except that the induced voltage from the rotor (again, dc current controlled rotating field), E_f, lags the terminal voltage by the torque angle zeta. The phasor diagram (again, on a per-phase basis) is presented in Figure 2.13b. Since the synchronous reactance is much greater than the internal resistance, the power-per-phase is given by Equation 2.6a. However, if one does consider this small amount of internal resistance, the full expression for power becomes Equation 2.6b. This small amount of resistance accentuates the shifting of the unity power factor point with load.

$$P(\text{per-phase}) = (V_t E_f/X_s)\sin\zeta \qquad (2.6a)$$

$$P(\text{per-phase}) = (V_t E_f/Z)\sin(\zeta - \alpha_z) + V_t^2 r/Z^2 \qquad (2.6b)$$

where $\alpha_z = \tan^{-1}(X_s/r)$.

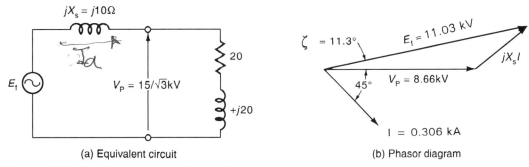

(a) Equivalent circuit   (b) Phasor diagram

Figure 2.12 Synchronous generator problem

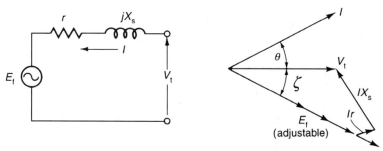

Figure 2.13 Typical per-phase equivalent circuit and diagram
for a synchronous machine

With reference to Figure 2.13b, since V_t is a constant and E_f is adjustable with a dc field current, the corresponding phasor directions (and lengths) may completely change. The resulting power factor (cos θ) characteristics are often represented by the well-known "V" curves (see Fig. 2.14).

It should be noted with reference to Figure 2.14 that the unity power factor locus (minimal points on the "V" curve) is not a vertical line but has a rather significant shift to the right. This shifting of the minimal point will be the cause of a changing power factor from unity for a change of external load.

As an example of using the adjustable power factor feature of synchronous motors, consider the following problem. Assume an existing plant, operating at full capacity, requires 100 kVA at 208 volts and has a lagging power factor of 0.8. Also assume the local power company charges a higher power rate to its commercial customers if their power factor falls below 0.8 (lagging). Now assume the plant manager has been asked to increase production by buying machinery that can be driven by 50 kW synchronous motor(s) that are adjustable from 0.71 leading to 0.71 lagging power factor (at full load). Neglecting efficiency, what is

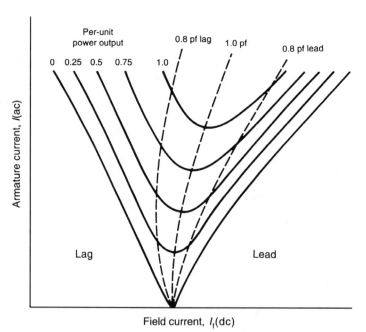

Figure 2.14 Typical "V" curves for an adjustable field synchronous
motor

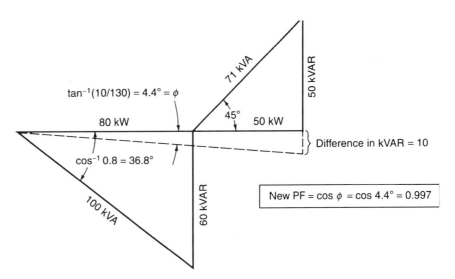

Figure 2.15 Power factor triangles

the kVA rating of the motor(s) and what would be the overall power factor if he bought the new machines and operated them at 0.71 leading power factor?

$$kVA_{new\ machines} = kW/PF = 50/0.71 = 70.4\ kVA$$

The total power triangle gives the new power factor (refer to Fig. 2.15).

It should also be noted that if the plant manager at a later date chose to increase power output of the new synchronous motors, he could do so by operating them at unity PF. For unity power factor, at essentially the same current as before, the power rating would be the same as the kVA rating. Thus, the new motor(s) could produce 71 kw or 21 kw of additional power!

More information concerning power factor correction techniques may be found in Chapter 3.

Induction Motors

For induction motors, usually with squirrel cage rotors, the speed is "almost" a function of frequency and the number of apparent pairs of poles. The stators of the synchronous machine and induction motor both produce a rotating magnetic vector that is a direct function of frequency and the number of apparent poles, whether multiphase or single phase. In contrast to the synchronous machine, in the case of the induction motor, torque can only be developed if there is a current due to an induced voltage in the squirrel cage rotor. Here, the rotor turns at a speed slightly less than synchronous speed, n_s. This is so because, for any current flow in the rotor, there must be an induced voltage in these rotor conductors. This voltage is caused by a relative motion between the rotating vector and the rotor. This is referred to as the slip, s. Slip is defined as

$$s = (n_s - n)/n_s \tag{2.7a}$$

$$n = (1 - s)n_s. \tag{2.7b}$$

Typical values of slip are in the range of 0.02–0.05 at full load; however, when the motor is running without driving a load and only enough torque is developed to overcome friction and windage losses, the slip drops to near zero.

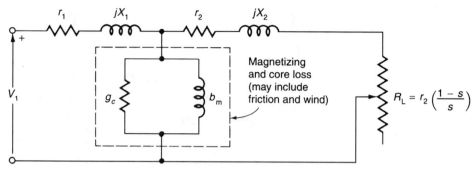

Figure 2.16 Induction motor equivalent circuit

As an example, assuming an induction motor has two pairs of poles and is rated at 60 Hz with a slip of 3 percent, the actual speed is easily determined. First, one finds the synchronous speed produced by the stator, which is given as

$$n_s = 60 \ f/\# \ PP = 60 \times 60/2 = 1{,}800 \text{ rpm}.$$

The rotor speed must be somewhat less to produce rated torque and may be found from Equation 2.7b as

$$n = (1 - s)n_s = (1 - 0.03)1{,}800 = 1{,}746 \text{ rpm}.$$

With this slip and other factors, these machines may be modeled and analyzed in detail for determining many different kinds of operating characteristic. The circuit model (see Fig. 2.16) is much like that of a transformer except for two major differences.

One difference comes from using a variable resistive load that is a function of slip (which also implies torque developed), given by $R_L = r_2(1 - s)/s$. (This represents an equivalent amount of mechanical power taken by the load.) This power may also include friction and windage loss; however, because this is a fairly constant value, it is often lumped in with the core loss. The other difference is that exciting current is not small and cannot be ignored. (This is so because of the extra mmf needed to produce the flux in the air gap between the rotor and stator.)

The induction motor, of course, takes electrical power and converts most of it to mechanical shaft power; and, when it does, it absorbs the electrical power. Here the current always lags the voltage, or is said to have a lagging power factor. As an example, consider a single-phase induction motor that is rated at 1 hp at 1180 rpm, 60 Hz, 240 V, with a known efficiency of 85 percent, operating at a lagging power factor of 0.8 (see Fig. 2.17). The question might be to find the current rating.

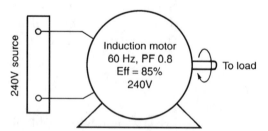

Figure 2.17 An induction motor operating at rated conditions

First, the power input is merely the power out divided by the efficiency,

$$P_{in} = P_{out}/\text{Eff} = 746W/0.85 = 877.4 \text{ watts,}$$

and the general power formula is

$$P = VI \times (PF) = 240\,I\,(0.8) = 877.4,$$

therefore $\quad\quad\quad\quad I = 4.57 \text{ A.}$

For this example, more information is given than is needed. However, because the speed and frequency are given, one could find the synchronous speed, the slip, etc. The synchronous speed is given for several different values of pair of poles by

$$n_s = 60\ f/\# \ PP = 3{,}600/\# \ PP, \ n_s = 3{,}600, 1{,}800, 1{,}200, \text{ etc.} \quad \textbf{(2.8)}$$

Because the rated speed (1,180 rpm) is slightly less than synchronous speed (obviously 1,200 rpm), the number of pairs of poles must be three. If the preceding problem were a three-phase induction motor, the main difference in obtaining the solution would be the general power formula of $P = \sqrt{3}\,V_L I_L (PF)$ and it wouldn't make any difference whether it was in delta or wye. While problems involving the slip and the number of pairs of poles for ac machines may not be on the examination, problems involving power factors and efficiencies may well be asked.

Series Motors

The series motor is sometimes called the universal motor, as it will operate on dc as well as ac. This kind of motor has brushes and an armature with the field pole windings in series with the armature. These units are used for both high-speed motors operating under a light load (*i.e.*, vacuum cleaners, high-speed fans, etc.) and high starting torques devices (electric drills, food mixers, etc.). The configuration of these motors is essentially the same as that given in Figure 2.8b, with the series winding but without the shunt winding. Problems involving series motors are usually limited in scope to concept only.

Specialized Motors

Specialized ac motors cover too many different devices to attempt to cover here. Nevertheless, one such device, an ac servo motor, will be included because of the likelihood of being referred to in a control system problem. However, if one is completely unfamiliar with this device, it is recommended that the reader skip this subject and use the study time elsewhere.

An ac servo motor is a two-phase controllable induction motor. These types of motors (at least small ones) generally can be used on a single-phase system with a second phase created artificially by a phase shifting capacitor in series with one of the motor windings. The servo two-phase motor is something like a single-phase induction motor (with its starting winding being left in the circuit) with the rotor being a squirrel cage type. In fact, the physical difference is that, as compared to a typical induction motor rotor, the servo motor usually is longer and with a smaller diameter; this is so because of the need to have as little as possible inertia, J, to assure fast stops and starts.

Recall that a single-phase induction motor has a pulsating magnetic vector rather than a rotating one; however, it is easily shown that a pulsating vector is exactly equivalent to two oppositely rotating vectors of half the magnitude of the original pulsating one. (This effect may be mathematically justified with the trig identity of $2\cos\omega t = e^{j\omega t} + e^{-j\omega t}$.) In other words, if one could isolate each half rotating vector by itself, the effect would be like two half-size induction motors running in opposite directions. The actual effect of combining these two machine means that at zero speed, the combination would develop a net torque of zero; however, if the rotor were started in either direction, the larger torque (off of zero speed) would cause the motor to "take off" in that direction. This is, of course, how one starts a single-phase machine. That is, put in a starting winding that is phased +/− 90° from the main winding to develop its own pulsating vector. This vector combines with the main pulsating one to produce a full rotating vector in the desired direction; then, after starting, the extra winding may be disconnected.

However, for a servo motor, the windings are matched and left in the circuit (See Fig. 2.18a). The main winding has a full voltage on it all of the time and is phased 90° (usually by a capacitor) with respect to the source voltage. The other winding is the control winding (which may come from a power amplifier at either 0° or 180°). These motors may be shown to develop a torque proportional to the magnitude of the control voltage (with the main winding voltage held constant); the voltage source is referred to as the carrier voltage. And, when working as a unit from the input ac voltage to the output shaft rotation, it is known as a "suppressed carrier system." Another factor needs to be considered: the speed should be kept well below that of the no-load speed. Much above half of this no-load speed, these motors tend to "single-phase" on the main winding alone! The torque-voltage-speed relationship (for the lower speeds) is shown in Figure 2.18b. It is easily demonstrated that the torque-voltage-speed equation may be found from this family of curves as

$$T_{developed}(s) = M\, s\, \Theta(s) + k\, V_{control}\,(s), \tag{2.9}$$

where M is the slope of the voltage "curves" (T/speed), and k is given by T_{rated}/V_{rated} at zero speed.

This portion on ac servo motors and resulting equation(s) may be important when reviewing the chapter on control systems (see Problem 5.6 in *Electrical Engineering: Problems & Solutions*).

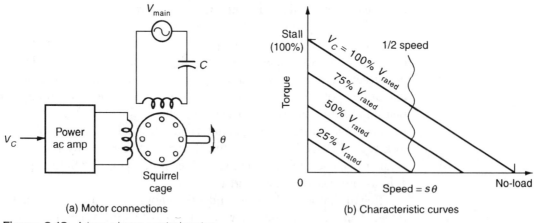

(a) Motor connections   (b) Characteristic curves

Figure 2.18 A two-phase control motor

RECOMMENDED REFERENCES

Chapman, *Electric Machinery Fundamentals,* 3rd Edition, McGraw-Hill, 1999.

El-Sharkawi, *Fundamentals of Electric Drives,* Brooks/Cole, 2000.

Fitzgerald, Kingsley and Umans, *Electrical Machinery*; McGraw-Hill, 2002. [For more on electrical machines and terms.]

Hambley, *Electrical Engineering Principles and Applications,* 3rd Edition, Prentice Hall, 2005 (Chapters 15, 16, and 17).

Kuo, *Automatic Control Systems*, Prentice Hall, any edition. [For more information on the servo motor.]

Matsch and Morgan, *Electromagnetic and Electromechanical Machines,* 3rd Edition, Wiley, 1986.

Nasar, *Electric Machines and Electromechanics, Schaum's Outline Series,* McGraw-Hill, 1981.

National Electrical Code (NEC), National Fire Protection Association, Inc., 2005.

Wildi, *Electrical Machines, Drives, and Power Systems,* 5th Edition, Prentice Hall, 2002.

Yamayee and Bala, *Electromechanical Energy Devices and Power Systems,* Wiley, 1994.

Power Distribution

OUTLINE

This chapter will include a very short review of three-phase theory. It will look
at power factor correction, unbalanced loads, and line faults, with a short review
of three-phase transmission theory, and transformer and synchronous equivalent
circuits as applicable to power transmission.

THREE-PHASE REVIEW

This discussion will start with a simplified explanation of the generation of a
three-phase voltage for a synchronous generator configuration (see Fig. 3.1). The
rotor is a controllable magnetic field pole that rotates at, say, 3,600 rpm (for 60 Hz).
(You may want to refresh your memory by reviewing the section on synchronous
machines in Chapter 2.) The output voltage will be a balanced three-phase voltage
source connected to any three-phase balanced load. The internal impedance of
the source is approximated by the synchronous reactance, jX_s, per-phase. Imagine
the source voltages as each being independent and connected in a wye configuration
with return leads. The voltages per-phase are as shown in Chapter 2 (Figs. 2.11
and 2.12, except V_T is now V_P). For purposes of this initial discussion, it is assumed
the voltages at the terminals are the source voltages and also the wye voltages
per-phase (actually this phase voltage will be shown to be the line voltage divided
by the square root of three).

Wye Connections

First consider that each source is connected to a resistive load as in Figure 3.2.
Each voltage source, or phase, is connected to each load (phase) by a line (notation
is important here). The bundle of return lines will shortly be called "the neutral."

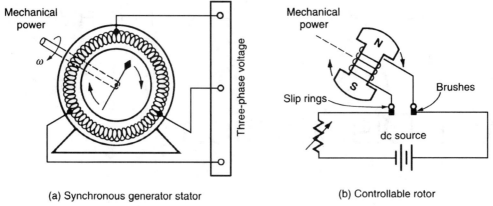

(a) Synchronous generator stator

(b) Controllable rotor

Figure 3.1 Synchronous alternator in generator mode

For example, assume each phase voltage is 100 volts and each resistor is 5 ohms. Then in Figure 3.2b, the current is V/R, or 20 amperes and the power dissipated in the resistor is $I^2 R$ or 2kW. Because all values are the same, the total power dissipated in all of the loads is merely the sum of the individual powers, or 6 kW. Here, the phase currents are the same as the currents in the lines. The currents in the bundle of return lines are, of course, also the same as the phase currents. Because the return currents are independent, the wires in the return bundle could be one single conductor, called the neutral (connect a', b', and c' to n). The current in the neutral wire is $I_n = I_a\angle 0° + I_b\angle -120° + I_c\angle +120° = 0$, as shown in Figure 3.3. Thus, since it carries zero current, there is no need for the neutral wire. However, in actual power utility circuits, the wire is normally left in, as a fourth wire, to carry any unbalanced currents. If one were to measure the voltage between any two lines, say from b to a, the voltage would be $V_{ba} = -V_{bn} + V_{na}$. This voltage is shown in Figure 3.4. Note that the resulting line voltage, V_{L1}, with respect to the phase voltage, V_a (implies V_{na}) is $\sqrt{3}$ times as large and leads the phase voltage by 30°. Of course the other two line voltages may be found in the same manner. It should be emphasized that for balanced loads, only one leg need actually be solved and the other two found by similarity. For a wye circuit,

$$I_{\text{Line}} = I_{\text{Phase}}, \ V_{\text{Line}} = \sqrt{3}V_{\text{Phase}}\angle +30° \tag{3.1}$$

Consider the following example problem: Assume a three-phase voltage source is available and has a measured line voltage of 173 volts (measured between any

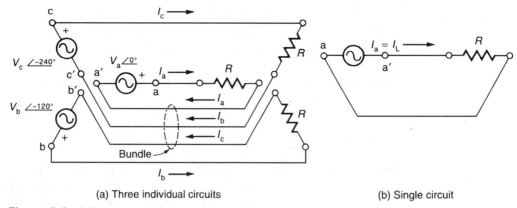

(a) Three individual circuits

(b) Single circuit

Figure 3.2 A three source system

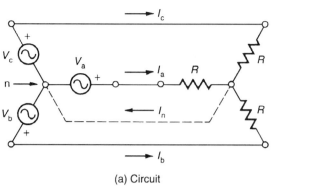

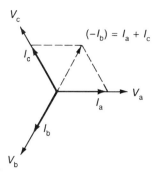

(a) Circuit

(b) Phasor diagram

Figure 3.3   A wye-wye three-phase circuit

two pins on a terminal board) and is connected to a balanced wye load of $Z = 5 + j5$ ohms in each leg (see Fig. 3.5). Find currents, voltages, and both power dissipated in the load and power taken from the source.

For this problem, the source comes from a three-terminal source that is not designated as either delta or wye; therefore, assume the easy case, that it comes from a wye source with a neutral wire. The phase voltage must be $V_{line}/\sqrt{3}$ and may be designated as reference. The problem is easily solved if one considers the individual branch as shown in Figure 3.5b.

$$V_p = 173/(\sqrt{3} \angle 0°) = 100 \angle 0° \text{ (assumed reference)}$$
$$I_P = (100 \angle 0°)/Z = 100 \angle 0°/(5 + j5) = (100 \angle 0°)/(7.07 \angle 45°)$$
$$= 14.1 \angle -45°$$
$$P_P = (I_P)^2 R = (14.1)^2 5 = 1 \text{ kW}$$
$$P_{tot} = \Sigma P_P = 3P_P = 3(1000) = 3 \text{ kW}$$

The total power taken from the source or dissipated by the full load is given by the following general three-phase formula (both delta and wye):

$$P_{tot} = \sqrt{3}\, V_L I_L \times \text{(power factor per phase)} \qquad (3.2)$$
$$P_{tot} = \sqrt{3} \times 173 \times 14.1 \cos 45° = 3 \text{ kW}.$$

Here again, the power factor is the cosine of the angle between the phase voltage and phase current (*i.e.,* the angle of the phase impedance). The phasor diagram is given in Figure 3.6. Recall that the phasor diagram is a "snapshot in time" and the whole diagram is continually rotating counterclockwise, so which phasor is chosen as reference is arbitrary.

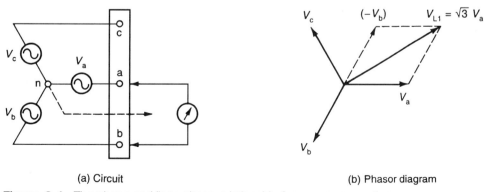

(a) Circuit

(b) Phasor diagram

Figure 3.4   The phase and line voltage relationship for a wye connection

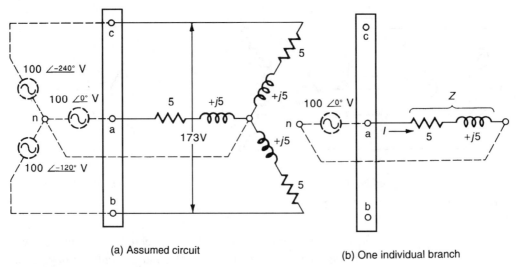

(a) Assumed circuit

(b) One individual branch

Figure 3.5 Balanced three-phase problem

Delta Connections

For a delta connection there is no neutral; therefore, the phase voltage across each leg of the delta must also be the line voltage (usually made reference) and each leg will be 120° out of phase with each other. The phase current(s) are merely the line voltage(s) divided by the phase impedance; the three currents, of course, are also 120° out of phase with each other (see Fig. 3.7).

As an example, the phase current $I_{Pab} = V_{Pab}/Z_{ab} = V_{Lab}/Z_{ab}$ is the same magnitude for each leg but phased 120° apart. The line current(s), on the other hand, is the phasor sum at each junction of the phase currents. For example, the line current, $I_{L1} = I_{P-ab} - I_{Pca}$, is easily shown to be equal to /3 times the phase current and lags by 30°. For a delta circuit,

$$V_{line} = V_{phase} \quad I_{line} = \sqrt{3}\, I_{phase} \angle -30° \qquad (3.3)$$

For another example, consider essentially the same problem as for the wye (Fig. 3.5) but connected as a delta. Again, assume the line voltage at a three-phase terminal board to be 173 volts and the phase impedance to be $5 + j5$ ohms. Here, the phase current within the delta is $I_P = (V_P)/Z$, or $I_P = (173 \angle 0°)/(5 + j5) = 24.5$ A. Note that for the same impedance and the same line voltage as that in a wye circuit, the current in the delta is $\sqrt{3}$ greater than that of the wye! The power per

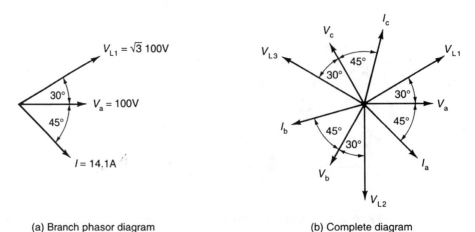

(a) Branch phasor diagram

(b) Complete diagram

Figure 3.6 Phasor diagram(s) for Figure 3.5

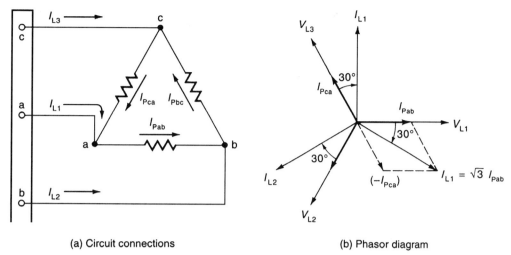

(a) Circuit connections   (b) Phasor diagram

Figure 3.7 The delta relationship

phase is found as before, $P - (I_p)^2 R$, or, $P = (24.5)^2\ 5 = 3$ kW, and thus the total power is 9 kW. Again, using the general power formula as a check, the power taken from the source is given by

$$P_{tot} = \sqrt{3}\ V_L I_L (PF), \tag{3.4}$$
$$P_{tot} = \sqrt{3} \times 173 \times (\sqrt{3} \times 24.5) \times 0.707 = 9 \text{ kW}.$$

Here, it is obvious that the total power is three times that of the wye connection for the same impedance values and the same line voltages.

Power Factor Correction

The transmission of power being delivered to a load that is not purely resistive may be very inefficient in terms of power line losses. If most of a load is inductive, the current rating of all parameters could be at an allowable maximum. Because the current lags the voltage by a large angle, the transmission line, any transformers in the path, and the source generator may be dissipating much more real power than necessary. The concept of correcting this problem is simple, but the details may be somewhat more complex.

Consider the following example problem. On a single-phase or a per-phase wye system, imagine a load of $3 + j4$ Ω requiring a 440 voltage source being fed by a power line with a total resistance of 0.5 Ω (see Fig. 3.8). For this particular problem, it will be assumed the power lines are purely resistive (although in actuality,

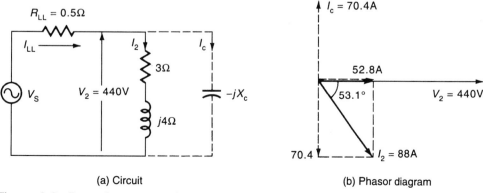

(a) Circuit   (b) Phasor diagram

Figure 3.8 Power factor correction problem

they may be more inductive than resistive). For the load, it is obvious that the current, I_2, is $440/(3 + j4) = 88.0 \angle -53.1°$ A, and power line loss is $P_{LL} = 88^2(0.5) = 3.87$ kW. And, the voltage source needs to be increased to $V_s = V_{LL} + V_2 = (88 \angle -53.1°)(0.5) + 440 \angle 0° = 468 \angle -4.3°$ V to support the 440 voltage at the load. By adding in parallel a capacitive reactive element with the load that would require a current just equal the imaginary portion of the load current, I_2, $(I_C = 88 \sin 53.1° = j70.4$ A), the new line current will be the phasor sum of the two currents. Because the two reactive components cancel each other, this sum is just equal to the real portion of the load current. Thus the line current is reduced $(I_{line} = I_2 + I_C = I_2 \cos 53.1° = 52.8 \angle 0°$ A). The new power line loss is also reduced, $P_{LL} = (52.8)^2(0.5) = 1.394$ kW, while the source voltage is decreased slightly, $V_s = IR_{LL} + V_2 = (52.8 \angle 0°)(0.5) + 440 \angle 0° = 466$ V. By adding an additional load, the power line loss has been greatly reduced.

For a more comprehensive three-phase problem, it is frequently easier to work with complex apparent power, $S = P +/- jQ$. For a three-phase *wye* system, the real and reactive power are given as

$$P_P = V_P I_L \cos \Theta = (V_L I_L \sqrt{3}) \cos \Theta = V_L^2/(3|Z_Y|) \cos \Theta \tag{3.5a}$$

$$Q_P = V_P I_L \sin \Theta = (V_L I_L \sqrt{3}) \sin \Theta = V_L^2/(3|Z_Y|) \sin \Theta. \tag{3.5b}$$

Then the total real and reactive power is three times the phase values, as

$$|S| = \sqrt{(3P_P)^2 + (3Q_P)^2} = 3\sqrt{P^2 + Q_P^2} = \sqrt{3} V_L I_L \tag{3.6}$$

and the total real and reactive powers are

$$P = 3P_P = |S| \cos \Theta \tag{3.7a}$$
$$Q = 3Q_P = |S| \sin \Theta. \tag{3.7b}$$

For a *balance delta load,* by using the delta-wye transformation equation it may be shown that $Z_Y = (1/3)Z_D$. Because the phase current for either a delta or wye is given as $I_P = V_P/(|Z|)$, for a delta, the line current is

$$I_L = \sqrt{3} V_L/(|Z_D|) \text{ A.} \tag{3.8}$$

From knowing that $I_L = V_L/(\sqrt{3} |Z_Y|)$, then it is clear *the current to a delta load is three times greater than to a wye load with each having the same phase impedance.* A short example follows.

For this three-phase example problem assume the impedance given by the previous problem is in each leg of a wye load (with the same 440 phase voltage). At the load terminals, another load in a delta configuration is connected in parallel with a capacitive reactance of $Z_D = 0 - j\ 15$ Ω (see Fig.3.9a).

The wye equivalent of the capacitor load is $Z_Y = (1/3)Z_D = -j5$ Ω; the equivalent impedance per leg of the wye (see Fig.3.9b) is $Z_{Yeq} = [5 \angle 53.1° \times 5 \angle -90°]/(3 + j4 - j5)] = 7.905 \angle -18.5°$ Ω.

The line current is given by $I_L = I_P = 440/Z_{eq} = 55.6 \angle 18.5°$ A. Because the line voltage at the terminals is $\sqrt{3} \times 440 = 726$ volts, the apparent power (from Eq. 3.6) is given as

$$|S| = \sqrt{3} \times 726 \times 55.6 = 69.9 \text{ kVa}$$
$$P = |S| \cos(-18.5°) = 69.9 \times 0.948 = 66.3 \text{ kW}$$
$$Q = |S| \sin(-18.5°) = (-) 69.9 \times 0.317 = (-) 22.2 \text{kVAR}$$

The line loss per-phase is $I^2 R_{LL} = 55.6^2 \times 0.5 = 1.55$ kW, compared to that of the wye branch of only $88^2 \times 0.5 = 3.87$ kW!

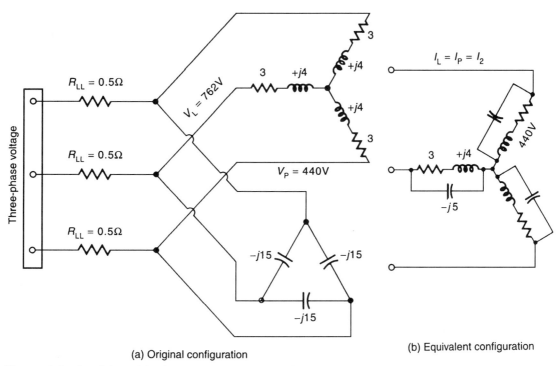

(a) Original configuration

(b) Equivalent configuration

Figure 3.9   Parallel combination of wye and delta loads

To find the optimal value of the capacitive reactance to reduce the line current to a minimum, one could use an apparent power triangle to find the necessary capacitive Q to cancel the inductive Q. Several of the more technical problems in Chapter 3 of *Electrical Engineering: Problems & Solutions* demonstrate the method.

Delta-Wye Conversion and Unbalanced Loads

For most balanced or unbalanced three-phase loads, one finds that either a delta or a wye configuration is more suitable for a particular kind of calculation. The conversion process is easy but sometimes tedious; however, having the "right form" will usually be worthwhile. As an example, assume a known delta balanced or unbalanced load is connected to a three-phase balanced source with a particular line impedance (see Fig 3.10).

If the load were a wye, rather than a delta, the line impedances would just add to each leg of the wye, yielding a new wye impedance, making any future calculations especially easy.

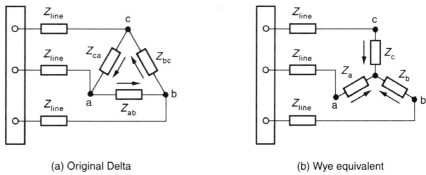

(a) Original Delta

(b) Wye equivalent

Figure 3.10   A delta–wye conversion for computational purposes

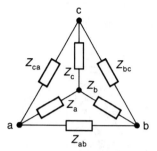

Figure 3.11 Delta–wye conversion notation

The elements for conversion are defined in Figure 3.11. The actual conversion formula from a delta to a wye is such that the impedance of the product of the adjacent delta elements divided by the sum of the wye elements is given in Equation 3.9a.

$$Z_a = \frac{Z_{ca}Z_{ab}}{Z_{ab} + Z_{bc} + Z_{ca}} \tag{3.9a}$$

On the other hand, sometimes a delta configuration is needed from a wye circuit (especially true for unbalanced loads). The delta impedance is the sum of the products of the wye elements divided by the opposite leg of the wye,

$$Z_{ab} = \frac{Z_aZ_b + Z_bZ_c + Z_cZ_a}{Z_c}. \tag{3.9b}$$

As an example, consider a wye configuration that is originally given as a balanced load of $5 + j5$, with the equivalent delta configuration needed. Here, the calculations of Equation 3.9b may be greatly simplified (because the load is balanced) as

$$Z_{ab} = Z_{bc} = Z_{ca} = Z_D = \frac{Z_aZ_b + Z_bZ_c + Z_cZ_a}{Z_c} = \frac{3(Z_Y)^2}{Z_Y}$$

$$Z_\Delta = 3Z_Y = 3(5 + j5) = 15 + j15\,\Omega.$$

Unbalanced Conversion

Suppose a wye configuration is unbalanced and a problem requires the computation of line currents. The problem, of course, is that for the wye, the phase voltages are not the same as the neutral phase shifts. Here, it is much easier to convert to an unbalanced delta with known line voltages. Assume, for the previous problem ($Z_Y = 5 + j5$), that one leg of the wye (say, Z_a) eventually develops a short in the inductance ($Z_a = 5 + j0$). Then, if one needs to know the delta equivalent, the calculations are somewhat more tedious and are computed to be

$$Z_{ab''} = \frac{5 \times 5\sqrt{2}\angle 45° + 5 \times 5(\sqrt{2})^2 \angle 90° + 5 \times 5\sqrt{2}\angle 45°}{5\sqrt{2}\angle 45°}$$

$$= \frac{50 + j100}{5\sqrt{2}\angle 45°} = \frac{11.8\angle 63.4°}{7.07\angle 45°} = 15.8\angle 18.4°$$

$$Z_{bc''} = \frac{5 \times 5\sqrt{2}\angle 45° + 5 \times (\sqrt{2})^2 \angle 90° + 5 \times 5\sqrt{2}\angle 45°}{5\angle 0°}$$

$$= \frac{50 + j100}{5\angle 0°} = \frac{11.8\angle 63.4°}{5\angle 0°} = 22.4\angle 63.4°$$

$$Z_{ca''} = \frac{5 \times 5\sqrt{2}\angle 45° + 5 \times 5(\sqrt{2})^2 \angle 90° + 5 \times 5\sqrt{2}\angle 45°}{5\sqrt{2}\angle 45°}$$

$$= \frac{50 + j100}{5\sqrt{2}\angle 45°} = \frac{11.8\angle 63.4°}{7.07\angle 45°} = 15.8\angle 18.4°.$$

To find the individual line currents for a balanced line voltage of 208 volts (assume the line resistance is zero), one uses the previously computed Z''_{deltas} rather than the wye configuration, as the phase voltages for the wye are not obvious. The computations for the delta phase currents are as follows:

$$I_{ab} = \frac{V_{ab}}{Z_{ab}} = \frac{208\angle 0°}{15.8\angle 18.4°} = 13.2\angle -18.4° = (12.5 - j4.17)A$$

$$I_{bc} = \frac{V_{bc}}{Z_{bc}} = \frac{208\angle 0°}{22.3\angle -63.4°} = 9.33\angle -56.6° = (5.4 - j7.79)A$$

$$I_{ca} = \frac{V_{ca}}{Z_{ca}} = \frac{208\angle 120°}{15.8\angle 18.4°} = 13.2\angle 101.6° = (-2.65 + j12.9)A$$

And, the line currents may then be found.

$I_{line-a} = I_{ab} - I_{ca} = (12.5 - j4.17) - (2.65 + j12.4) = 15.2 - j16.6 = 22.5\angle -47.5° \text{ A.}$
$I_{line-b} = I_{bc} - I_{ab} = (5.1 - j7.79) - (12.5 - j4.17) = -7.36 - j3.62 = 8.20\angle -153.2° \text{ A.}$
$I_{line-c} = I_{ca} - I_{bc} = (-2.65 - j12.9) - (5.14 - j7.79) = -7.79 + j20.7 = 22.1\angle 110.6° \text{ A.}$

The reader may wish to work the same problem with a 1.0 ohm line impedance (in series with each leg of the unbalanced wye) before converting to a delta.

Symmetrical Components

One method of solving unbalanced three-phase problems, especially for larger complex systems, is that of using symmetrical components. This is a procedure of reducing unbalanced vectors into three symmetrical components. The advantage of obtaining these symmetrical components is that they are balanced and therefore easy to solve.

The set of equations is obtained by considering a positive balanced sequence of phasors (say, abc), a negative sequence (say a, c, b), and finally a zero sequence. If one uses the subscript, 1, as the positive sequence, 2, as the negative sequence, and, 0, for the zero sequence, then for each of the unbalanced phases (say, for an unbalanced voltage), one may reduce each phasor to a set of three balanced set of phasors, as

$$V_a = V_{a1} + V_{a2} + V_{a0} \tag{3.10a}$$
$$V_b = V_{b1} + V_{b2} + V_{b0} \tag{3.10b}$$
$$V_c = V_{c1} + V_{c2} + V_{c0} \tag{3.10c}$$

Here, each set of phasors is related by 120° as follows:

$$V_{a1} = V_{b1}\angle 120° = V_{c1}\angle -120° \tag{3.11a}$$
$$V_{a2} = V_{b2}\angle -120° = V_{c2}\angle 120° \tag{3.11b}$$
$$V_{a0} = V_{b0} = V_{c0} \tag{3.11c}$$

Equations 3.11a, b, c may then be written as

$$V_a = V_{a1} + V_{a2} + V_{a0} \tag{3.12a}$$
$$V_b = V_{a1}\angle -120° + V_{a2}\angle 120° + V_{ao} \tag{3.12b}$$
$$V_c = V_{a1}\angle 120° + V_{a2}\angle -120° + V_{a0} \tag{3.12c}$$

At this point it should be pointed out that some textbooks present this information in a slightly different fashion. That is, just as "j" represents a phase shift of 90°

(and −1 for a phase shift of 180°), one could arbitrarily choose a symbol, say "a", to represent a phase shift of 120°. This symbolic representation is then given as (only a partial list)

$$a = 1\angle120° = -0.5 + j0.866 \tag{3.13a}$$
$$a^2 = 1\angle240° = -0.5 - j0.866 \tag{3.13b}$$
$$a^3 = 1 + j0' \tag{3.13c}$$
$$1 + a = 0.5 + j0.866 = 1\angle60°3 = -a^2 \tag{3.13d}$$
$$1 + a + a^2 = 0 + j0. \tag{3.13e}$$

It is easily shown that Equations 3.13a,b,c may be written as

$$V_{a1} = 1/3(V_a + V_b\angle120° + V_c\angle-120° \text{ or } = 1/3(V_a + aV_b + a^2V_c) \tag{3.14a}$$
$$V_{a2} = 1/3(V_a + V_b\angle-120° + V_c\angle120° \text{ or } = 1/3 (V_a + a^2V_b + aV_c) \tag{3.14b}$$
$$V_{a0} = 1/3(V_a + V_b + V_c) \tag{3.14c}$$

Consider the previous example of a three-phase, 208 line voltage connected to a balanced delta load of $15 + j15$ Ω for each leg. Also assume a lossless line with $Z_{line} = 0$ Ω. For this configuration the line currents were 17.0 A and the power taken from the source was 4.32 kW. Now, however, assume one of the lines (say line c) had an open circuit (see Fig. 3.12) to completely unbalance the system; it essentially becomes a single-phase circuit. The load, then, becomes $(15 + j15)\|(30 + j30) = 14.1\angle45°$ Ω; the line current is $208/14.1 = 14.75$ A (at $\angle-45°$, but, for ease of calculation, it will be called reference, $\angle0°$). By making use of the preceding equations, the symmetrical components will be found from these known quantities ($I_a = 14.75\angle0°$, $I_b = 14.75\angle180°$, $I_c = 0$).

$$I_{a1} = 1/3(14.75\angle0° + 14.75\angle300° + 0) = 1/3(22.1 - j12.8) = 8.51\angle-30°$$
$$I_{a2} = 1/3(14.75\angle0° + 14.75\angle420° + 0) = 1/3(22.1 + j12.8) = 8.51\angle + 30°$$
$$I_0 = 0.$$

By similar calculations, the other symmetrical components are calculated to be

$$I_{b1} = 8.51\angle-150°, \quad I_{b2} = 8.51\angle + 150°, \quad I_{b0} = 0$$

and

$$I_{c1} = 8.51\angle90°, \quad I_{c2} = 8.51\angle-90°, \quad I_{c0} = 0.$$

Note that line c appears to be carrying a current; however, the phasor sum of I_{c1} and I_{c2} is actually zero. And, the line current I_a checks to be 14.75 A, and gives a new lower load power of only 2.176 kW. Other problems will show the power of the system.

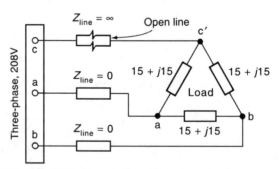

Figure 3.12   An unbalanced system due to an open line

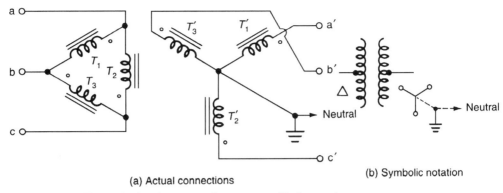

(a) Actual connections

(b) Symbolic notation

Figure 3.13 Three single-phase transformers used in three-phase

POWER TRANSMISSION

Most "bulk" power is transmitted in a three-phase configuration. The voltage levels are changed by banks of transformers, these being banks of three single-phase transformers, or one three-phase unit (or, in certain instances, a bank of two single-phase transformers connected as an open delta). One common case is for power, coming from a relatively low voltage generating delta source, to be converted to a higher voltage wye configuration with a neutral or ground wire (see Fig. 3.13). This power would go over a transmission line to a distribution point. The notation for this conversion operation is reduced to that of Figure 3.13b.

As an example, consider a three-phase source voltage of 12 kV that needs to be converted to 208 (actually, 207.8) kV line voltage along with a neutral wire to feed a transmission line. This transformation may be done by three single-phase 1:10 transformers. The individual phase voltage across each primary is 12 kV (in delta) and that of the secondary (in wye) is 120 kV (phase), while the wye's line voltage is $\sqrt{3}$ times its phase voltage, $V_L = \sqrt{3} \times 120 = 208$ kV. If the KVA rating of each transformer is 1000, the rated phase current on the 12 kV side is 1000/12 or 83.3 A and on 120 kV side is 8.33 A, and the total rating of the bank is 3 MVA (see Fig. 3.14).

As a check, the line currents to the delta are $\sqrt{3} \times 83.3 = 144.3$ A, then total $kVA_{in} = \sqrt{3} \, V_L I_L = \sqrt{3} \times 12 \times 10^3 \times 144.3 = 3,000$ kVA and the kVA_o is $= \sqrt{3} \times 208 \times 10^3 \times 8.33 = 3,000$ kVA.

For a power system, frequently the analysis is done on a per-phase basis using per unit values (see the section "Transformer per Unit Values" in Chapter 2). As an example, while the load could be as high 1,000 kVA (per-phase for a wye), assume it is only delivering 800 kW to a unity power factor load. Then the pu for the load leg and transformer leg is 800/1,000 = 0.8 pu's. The current legs are also 0.8 pu's while the voltages are 120 kv/120 kv = 1.0 pu's. Of course if one wanted to establish the line values as the basis, the ratios would still be the same.

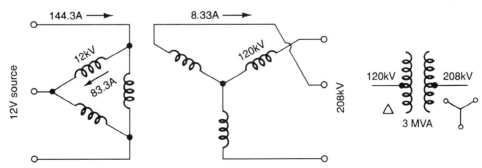

Figure 3.14 A three MVA three-phase system

Transients in Power Systems

Under normal operating conditions the synchronous reactance, neglecting the internal resistance, is frequently sufficient for performing steady state analysis of a synchronous generator with a connected load (see the section "Synchronous Machines" in Chapter 2). But under sudden load changes or short circuit conditions, other effects, called subreactance and/or transient reactance, should also be considered.

For the synchronous machines, two different type rotors are normally used. One is the cylindrical rotor (also called round rotor), usually used if the prime mover is a high-speed device (say 1,800 or 3,600 rpm), such as a steam turbine. The other is the salient pole type, usually used if the prime mover is slower and the number of poles is high (say 72 poles at 100 rpm). The normal synchronous reactance, $x_{s,}$ is thought of as $x_d + x_1$, where x_d is the direct reactance and x_1 is the leakage reactance (usually small and neglected). Again, for steady state operation, the flux at the pole is a maximum because the air gap is at its shortest, while the flux in quadrature (that is, for the two-reactance theory, the flux located half distance between the poles) is small because the air gap is a maximum. This flux causes a quadrature reactance, x_q, which is a minimum. However, under transient or short circuit conditions, a sudden large current is present in the stator windings and could cause a large current to be induced in the rotor, especially so for the salient pole machines. The effect of this induced rotor current is to change the armature mmf from the rotor so as to influence a transient current; a typical short circuit (or partial short circuit) current might appear as shown in Figure 3.15a. The high current immediately after a short circuit is applied may be designated by three time periods: subtransient period, lasting only one or two cycles with a current, i''; the transient period, with a current i'; and the normally steady state period or performance.

There are several other effects such as a changing of the "dc" field current with a transient, and the physical damping effect due to the design of the armature and rotor. However, in this explanation, these effects will not be considered individually. Instead, the lumping of some of these effects will be included into the direct and quadrature reactances. A skewing (or a decaying dc term) along

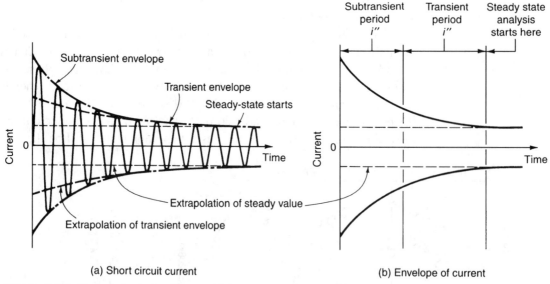

(a) Short circuit current   (b) Envelope of current

Figure 3.15 Typical short-circuit current for a synchronous machine

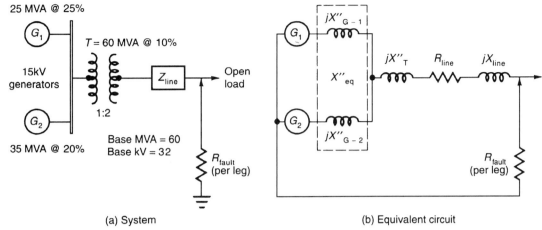

Figure 3.16 Subtransient problem arrangement

with the ac wave is possible depending on where in the stator current cycle the short circuit occurs, but will be ignored here. This effect could result in harmonics (also frequently ignored) of the current during the transient period(s).

During a current surge (usually due to a line fault) the net transient may be analyzed by defining a subtransient reactance as x_d'' (for the first part of the curve in Figure 3.15), a transient reactance as x_d' (as the mid part of the curve), and x_d (for the steady state portion of the analysis). A short example will follow.

By way of explanation, this short example will make use of per unit values in the analysis (it is recommended that one review this concept before proceeding). Assume that two three-phase generators (same voltage) are tied together to feed a load (for this example, an open load) through a step-up transformer as shown in Figure 3.16a. The transmission line impedance from the transformer to an open load is known to be $2 + j10\ \Omega$. Now assume that a very low fault resistance of $0.5\ \Omega$ (balanced in each leg of a wye) to ground occurs and that one needs to know the subtransient fault current. First it is necessary to convert the individual subtransient reactance of each generator to the base reactance of the system,

$$x_{G\text{-}1}'' = 0.25(60\text{MVA}/25\text{MVA}) = 0.6 \text{ pu},$$
$$x_{G\text{-}2}'' = 0.20(60\text{MVA}/35\text{MVA}) = 0.343 \text{ pu}.$$

The equivalent circuit of Figure 3.16b shows that the subtransient reactance has an equivalent of

$$x_{G\text{-}eq}'' = x_{G\text{-}1}'' \| x_{G\text{-}2}'' = [0.6 \times 0.343/(0.6 + 0.343)] = 0.2182 \text{ pu}.$$

And, for the transformer reactance,

$$x_{\text{transformer}}'' = (60 \text{ MVA}/60 \text{ MVA}) \times 0.1 = 0.1 \text{ pu}.$$

To convert the actual line impedance and the ground fault resistance to per-unit values, one must first find the base impedance,

$$\mathbf{Z}_{\text{base}} = V_{\text{base}}/I_{\text{base}} = V_{\text{base}}/(VA_{\text{base}}/V_{\text{base}}) = V_{\text{base}}{}^2/VA_{\text{base}},$$
$$\mathbf{Z}_{\text{line}}(pu) = Z_{\text{actual}}/Z_{\text{base}} = (2 + j10)[60 \times 10^6/(32 \times 10^3)^2]$$
$$= (2 + j10)(0.0586) = 0.1172 + j0.586 \text{ pu}$$
$$\mathbf{Z}_{\text{fault}}(pu) = (0.5)[0.0586] = 0.0293 \text{ pu}.$$

The subtransient current is computed by referring to Figure 3.17.

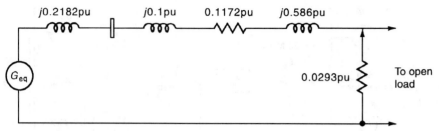

Figure 3.17   Converted equivalent circuit

Thus, the total per-unit impedance to ground is given as

$$\sum \mathbf{Z}(\text{pu}) = (0.1172 + 0.0293) + j(0.2182 + 0.1 + 0.586)$$
$$= 0.1465 + j0.9046 \text{ pu} \quad \Rightarrow \quad 0.9164\angle 80.8° \text{ pu}.$$

The modified circuit (including the fault resistance) MVA becomes

$$\text{MVA}_{\text{modified}} = \text{MVA}_{\text{orig}}/Z_{\text{pu}} = 60 \text{ MVA}/0.9164 \text{ pu} = 65.47 \text{ MVA},$$

giving the fault current as

$$I_{\text{fault}} = \text{MVA}_{\text{modified}}/(\sqrt{3}\,\text{kV}) = 65.47 \times 10^6/(\sqrt{3} \times 32 \times 10^3) = 1182 \text{ A}.$$

In addition to this text's discussion, an excellent and concise review of transients and subtransients, including practice problems, can be found in Nasar's *Electric Power Systems*.

Transmission Lines

For actual transmission lines with losses, the lines themselves should be considered along with their length. First, however, some mention of a power line equivalent circuit should be made.

Consider a three-wire transmission line in a triangular shape with each wire having a radius of R and a wire spacing between the conductors of D (see Fig. 3.18a).

The inductance per unit length, L', of each wire (from transmission line theory) is given by Equation 3.15.

$$L' = 2 \times 10^{-7} \ln D/R \text{ henrys/meter} \tag{3.15a}$$
$$L' = 0.741 \times 10^{-3} \log D/R \text{ henrys/mile} \tag{3.15b}$$

The capacitance of a three-phase transmission line, for Figure 3.18b, from line to neutral is as in Equation 3.16,

$$C' = (38.9 \times 10^{-9})/\log(D/R) \text{ farads to neutral/mile}. \tag{3.16}$$

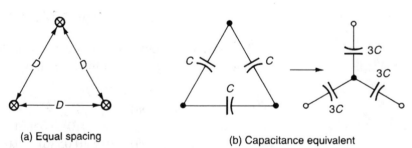

(a) Equal spacing   (b) Capacitance equivalent

Figure 3.18   Three-phase lines with an equal spacing, D

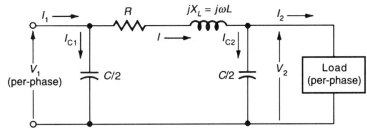

Figure 3.19   Equivalent circuit and load on a per-phase basis

These equations for L' and C' are for idealized conditions, and, in practice, a number of modifications need to be made which are beyond the scope of this review.

Most transmission lines (assuming 60 Hz) are relatively short[*] (less than fifty miles) compared to the wavelength being considered, and a lumped parameter π equivalent circuit (see Fig. 3.19) will give a reasonable representation. In many problems, the inductive reactive term is much greater than either R or C term and only the series inductance is needed to represent the line on a per-phase basis. An example problem follows.

A 12 kV three-phase system 50 km long has three conductors and delivers power to a total complex load of $20 + j5$ MVA (or $20/3 + j5/3$ MVA per phase) at a required 12 kV line voltage.

The power line parameters per-phase are found to be

$$\text{Resistance} = 0.0\ 2 \text{ ohms/km}$$
$$\text{Inductance} = 0.212 \text{ mh/km}$$
$$\text{Capacitance} = 10 \times 10^{-9} \text{ F/km (each).}$$

The power line parameters are then found for the 50 km length to be $R = 1.0\ \Omega$, $L = 10.6$ mh, $C = 500 \times 10^{-9}$, giving $X_L = 2\pi 60 \times 10.6 \times 10^{-3} = 4.0\ \Omega$, and $Y_C = 2\pi 60 \times 500 \times 10^{-9} = 1.885 \times 10^{-4}$ $1/\Omega$. The current at the receiving end is found from the per-phase complex power divided by the phase voltage $I_2 = (20/3 + j5/3$ MVA$)/(12/3$ kV$) = 0.962 + j0.240$ kA, the current through the right shunt capacitance is $I_C = (12/3$ kV$)\,(j1.885 \times 10^{-4}) = j0.001306$ kA. The voltage needed at V_1 will be the phasor sum of the V_2 plus the voltage drop across the series $R + jX_L$, where the current, I, through this series impedance is $I = I_2 + I_C$. However, since $I_C \ll I_2$, the current $I = I_2$, and the capacitance may be ignored in this particular problem,

$$V_2 = I_c / Y_c = \frac{0.001306\,kA}{1.885 \times 10^{-4}}$$

$$V_1 = (R + jX_L)I + V_2$$
$$= (1 + j4)(0.963 + j0.242) \times 10^3 + 6.93 \times 10^3 = 7.99\angle 29.8° \text{ kV.}$$

Or, the new source line voltage needed to support the load voltage is now $\sqrt{3}V_1 \angle + 30° = 13.8/59.8°$ kV.

---

[*]Depending on the degree of accuracy and method of solution used, problems are frequently set up for various transmission line lengths; the usual accepted method is to define three lengths. Short lines are defined as 50 miles or less, where shunt effects are usually neglected; medium lengths are between 50 and 150 miles, where shunt capacinces (lumped parameter) are normally considered; and, long lines are greater than 150 miles and the line is represented as a distributed parameter type line.

RECOMMENDED REFERENCES

Bergen and Vittal, *Power Systems Analysis,* 2nd Edition, Prentice Hall, 2000.

Bosela, *Electrical Systems Design,* Prentice Hall, 2003.

Chapman, *Electric Machinery and Power System Fundamentals,* McGraw-Hill, 2002.

Elgerd, *Basic Electric Power Engineering,* Addison-Wesley, any edition.

El-Hawary, *Electrical Energy Systems,* CRC Press, 2000.

Glover and Sarma, *Power System Analysis and Design,* 3rd Edition, Brooks/Cole, 2002.

Grainger and Stevenson, *Power System Analysis,* McGraw-Hill, 1994.

Nasar, *Electric Power Systems, Schaum's Outline Series,* McGraw-Hill, 1990.

National Electrical Code (NEC), National Fire Protection Association, Inc., Quincy, MA.

Sadat, *Power System Analysis,* McGraw-Hill, 1999.

Stevenson, William D. Jr. *Elements of Power System Analysis,* McGraw-Hill, p. 30–38.

Electronics

This chapter presents three areas for review: diodes, transistors, and integrated circuits (mostly operational amplifiers). These subjects start with a short review of basic concepts, along with short examples, and progress to a level expected of an electrical engineer. When possible, a linear model is used so as to take advantage of all circuit reduction techniques of analysis. Because the area of electronics is so broad, only portions of theory can be presented. The reader is urged to have a good reference text(s) available when proceeding through this review.

It is important, especially in the field of electronics, to pay attention to subscript notation. If the symbol is uppercase it implies a dc or average value if the subscript is also uppercase; if the subscript is lowercase, it may imply an rms or a function of frequency (either in the s-domain or the $j\omega$-domain). On the other hand, if the symbol is lowercase, and the subscript is uppercase, a time-varying function with a dc value is implied. If the subscript is also lowercase, the implication is a small signal time-varying quantity without a dc component. When working examination problems, the reader is urged to take into context what seems reasonable, as many notational mistakes may inadvertently be shown in a text or on an examination problem.

DIODES

The solid state theory of diode operation depends on the particular kind of diode being considered. For example, for the junction diode, knowledge of solid state theory and the behavior of majority and minority carriers in the presence of an

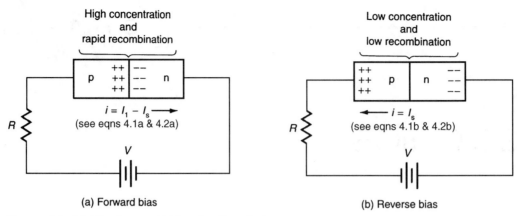

Figure 4.1 Mobile charges inside a junction diode

electric field is necessary. But, when treating the diode from its external characteristics, the full theory is not essential.

For a p-n junction diode, a very simplistic explanation of the operation at the junction between a "p" and an "n" type semiconductor is that in each type material there are many "free" charges available; these are holes (p-positive), and electrons (n-negative), respectively. If the diode is biased in favor of forcing positive charges near the boundary (the positive voltage being connected to the p material and the negative terminal to the n material), a rapid recombination of the charges takes place (see Fig. 4.1a); this is the direction of easy current flow. On the other hand, if reverse biased (see Fig.4.1.b), the free charges are attracted toward their bias polarities, leaving a dearth of charges at the junction, for very little recombination and almost no current flows. When the bias is in the forward direction, the voltage necessary to cause the charges to recombine is fairly small but still could be significant (see Fig. 4.3a). The voltage is near a half volt (approximately 0.3 to 0.4 volts for germanium and near 0.7 volts for silicon).

If one were to consider the diode current made up of two components, an injection current, I_i, that is a function of junction potential, and a reverse saturation current, I_s, that is not sensitive to junction potential, then the total current may be given as

$$i = I_i - I_s = Ae^{qV/kT} - I_s. \tag{4.1a}$$

Here the injection process is related to the statistical probability of electrons and holes having sufficient energy for recombination. And, with an increasing operating temperature and/or a forward bias voltage, the probability increases. For an open circuit condition with $i = V = 0$, it is clear that the constant A reduces to I_s. At room temperature (20°C) and for a germanium diode, e/kT is approximately 40 V^{-1}, and Equation 4.1a may be written as

$$i = I_s(e^{40V} - 1). \tag{4.1b}$$

For an introductory example consider a diode (germanium) whose reverse current at −1 volts is −50 μA. At room temperature, it is desired to find the currents for the forward and reverse voltages of 0.2 and −0.2 volts. The reverse current is given as −50 μA, which is effectively equal to $-I_s$. At the reverse bias voltage of −0.2, $e^{40(-0.2)}$ is almost zero. Thus $i = -I_s = -50$ μA. And, at 0.2 volts, $e^{40(0.2)} = 3,000$; thus $i = 50(3,000 - 1) \times 10^{-6} = 150$ mA.

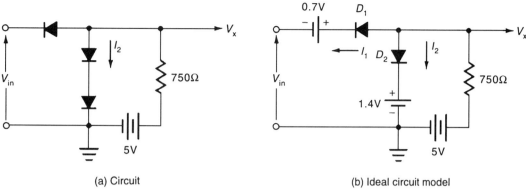

(a) Circuit

(b) Ideal circuit model

Figure 4.2 A diode switching circuit for the example problem

It may be observed that Equation 4.1b reduces to an even simpler form for a forward bias voltage greater than +0.1 and a reverse bias voltage of less than −0.1 volts,

$$i = I_s e^{40V} \text{ (forward bias, } V > 0.1 \text{ V)} \tag{4.2a}$$

$$i = -I_s \text{ (reverse bias, } V < -0.1 \text{ V).} \tag{4.2b}$$

This equation "fits" well for both germanium and silicon at higher currents, but for lower or normal currents, a better "fit" for silicon is

$$i = Ae^{eV/2kT} - I_s = I_S(e^{20V} - 1) \tag{4.3}$$

For another example, consider the silicon diode as a circuit element and its ideal circuit model (with a diode voltage of 0.7 volts) of Figure 4.2a, b. This is a switching circuit and it is interesting to observe the current, I_2, as a function of the input voltage, V_{in}.

For the ideal model, the actual diodes are replaced with an ideal diode in series with its threshold voltage of 0.7 volts (each). Note that if V_x is > 1.4 volts, then I_2 is in a direction of easy flow to ground. The voltage, V_x, cannot exceed $V_{in} + 0.7$ volts without D_1 being forward biased and the current, I_1, is in the direction of easy flow to the V_{in} source. Thus, if V_{in} is between 0 and 0.7 volts, the current from the 5 volt source becomes I_1 and $I_2 = 0$; on the other hand if $V_{in} > 0.7$ volts, diode D_2 is forward biased and I_2 is given as

$$I_2 = V/R = (5 - 1.4)/750 = 4.8 \text{ mA.}$$

A very small change in V_{in} can cause I_2 to jump from 0 to 4.8 mA.

The preceding equations are approximate and apply only if the reverse bias is not too great. For a larger reverse bias, another effect, referred to as the reverse breakdown voltage or the avalanche breakdown of the diode, takes place. However, the avalanche effect occurs at a voltage somewhat higher than the Zener breakdown voltage. Diodes designed to operate this region are called Zener diodes (see Fig. 4.3a, b).

In the reverse direction, the current is essentially zero until breakdown voltage is reached. Caution is urged when designing circuits using these diodes so as not to exceed the power or current Zener ratings.

In addition to the regions for the various operation of the diode already discussed, there is at least one more special-purpose operation that will be addressed. This special-purpose application makes use of the region between zero current and the Zener region; that is, when an extremely small current is flowing

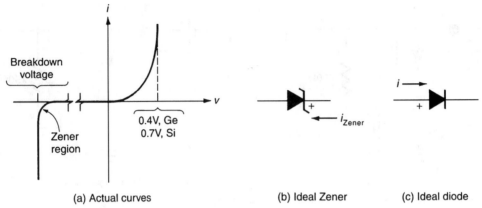

(a) Actual curves   (b) Ideal Zener   (c) Ideal diode

Figure 4.3 Junction diode voltage-current characteristics

in the reverse direction, the diode has the characteristics of a capacitance. The reverse voltage causes a depletion region at the physical junction and has a dearth of carriers (like an insulator) but with fixed positive and negative charges. Diodes made especially to enhance this capacitance property are known as varactors or varicaps. One typical application of a varactor is for the variable capacitive element used in automatic frequency control circuits in FM radio. This (reverse) capacitance, C_R, decreases with reverse voltage and is considered controllable with voltage (see Fig. 4.4).

The reverse capacitance is given approximately by

$$C_R = K/(V_o - v_R)^m \tag{4.4}$$

where (C_R typically may vary from 1 to 500 pF)

V_o = the depletion layer voltage with 0 applied voltage (v_R).

v_R = applied voltage at the terminals.

K = a constant that is a function of the junction area and the diode doping.

m = a constant that is a function of the distribution of impurities near the junction (typical values range from 0.3 to 4).

If larger capacitances are needed there is still another diode region of operation that is used; this is just in the forward direction but well before the threshold voltage is reached. Here the depletion region has a high concentration of minority carries and the effect is to produce a diffusion capacitance, C_D. The value of C_D is a function of the forward current; values of C_D are typically much greater than those of the C_R's.

Many good references are available for diode theory and electronic circuits without being overloaded with solid state physics. A particularly good one is the text by Sendra and Smith.

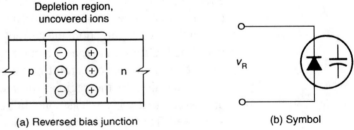

(a) Reversed bias junction   (b) Symbol

Figure 4.4 A varactor junction diode (circled charges indicate fixed ions)

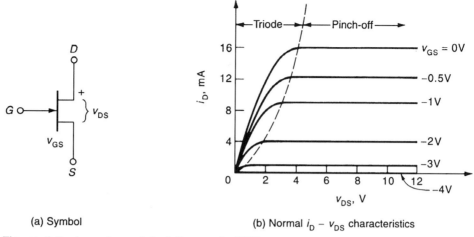

(a) Symbol   (b) Normal $i_D - v_{DS}$ characteristics

Figure 4.5 An n-channel depletion mode JFET

TRANSISTORS

Despite the diversity of transistors being developed, one would expect the kinds of problems on the professional examination to be limited to either the bipolar junction transistor (BJT) family or the field-effect transistor (FET) family. In general one may consider the BJT family to be current controlled while the FET family is voltage controlled.

These controlling signal sources, either current or voltage, allow for diverse applications from signal amplification to switching circuits. Methods of design using these devices range from graphical load-line analysis (especially when operating in a nonlinear region, over large signal swings) to small signal and h-parameter techniques of analysis.

Field-effect Transistors

Within the FET family, a further breakdown of the technology yields two sub-families, the junction field-effect transistors (JFETs) (see Figure 4.5), and the metal-oxide transistors (MOSFETs).

There is a still further breakdown within the MOSFET family, which acts in either an enhancement or depletion mode of operation (see Fig.4.6); the curves[*] of Figure 4.6b are measured dc (static) characteristics.

The n-channel enhancement mode (Fig. 4.6) offers the best performance because, in the n-material, the higher mobility of the electrons leads to faster switching times. Because the metal oxide layer insulates from the substrate of material, these devices are sometimes called insulated-gate FETs, or IGFETs (here typical resistances are in the order of 10^{12} to 10^{15} ohms). Just after the bend in the knee (the dashed line) in Figure 4.6b, the curves have almost constant current characteristics. This is called the active region, while to the left of the dashed line, is the ohmic or triode region.

---

[*]Typical curves as presented in Carlson and Gisser, *Electrical Engineering Concepts and Applications,* Addison-Wesley, 1990, p. 286.

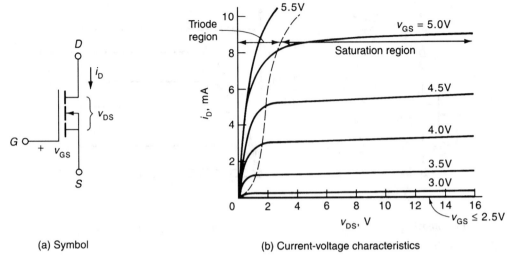

(a) Symbol

(b) Current-voltage characteristics

Figure 4.6 A MOSFET n-channel enhancement mode

The MOSFET (enhancement mode), in addition to the i_D vs v_{DS} characteristics of Figure 4.6b, has an $i_{DS} - v_{GS}$ relationship in the triode region (where $V_{DS} < V_{DS,sat}$) is given by Equation 4.5,

$$i_D = K[2(v_{GS} - V_T)v_{DS} - v_{DS}^2], \; v_{DS} < v_{DS,sat} \qquad \textbf{(4.5a)}$$

And, for the saturation region (where $v_{DS} > v_{DS,sat}$), it can be shown that,

$$v_{DS,sat} = (v_{GS} - V_T). \qquad \textbf{(4.5b)}$$

Let $v_{DS} = v_{DS,sat}$ in Equation 4.5a. Then, substituting Equation 4.5b into Equation 4.5a yields

$$i_D = K(v_{GS} - V_T)^2, \; v_{DS} > v_{DS,sat} \qquad \textbf{(4.5c)}$$

The V_T is the turn-on or threshold voltage (usually specified by the manufacturer and ranging from 1 to 3 volts), and, K is the relating parameter, typically near 0.3 mA/V^2.

An example problem for finding the relating parameter K follows. Assume for the characteristic curves of Figure 4.6 (with a $V_T = 2.5$ V) that $I_{DSS} = 2$ ma at a $v_{GS} = 5$ V; then,

$$i_D = K(v_{GS} - V_T)^2 = K(5 - 2.5)^2 = 2.0 \text{ ma},$$
$$K = I_{DSS}/(V_T)^2 = 2 \times 10^{-3} \text{ A}/(2.5)^2 = 0.32 \text{ mA/V}^2.$$

(More on this relationship later.)

Also, as an extension problem, find the load resistance, R_D, for an operating quiescent point at a current of 3.2 ma and a v_{DS} of 8.5 V for the circuit of Figure 4.7.

For finding the operating quiescent point and the corresponding load-line for a $V_{DD} = 15$ volts and $R_G = 0.5 \times 10^8 \; \Omega$, also determine R_D for an operation point of $v_{DS} = 8.5$ volts and an i_D of 3.2 mA. To find the resistive load, R_D, which is the slope of a load-line superimposed on the characteristic $i_D - v_{DS}$ curves, one determines R_D from

$$R_D = (V_{DD} - v_{DS})/i_D = (15 - 8.5)/(3.2 \times 10^{-3}) = 2.03 \times 10^3 = 2 \text{ k}\Omega.$$

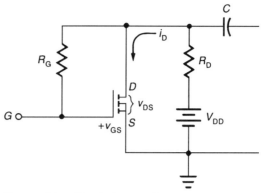

Figure 4.7 Circuit diagram for example problem

For linear operation, the added constraints are

$$|v_{DS}| < (v_{GS} - V_T), \ v_{GS} > V_T. \quad\quad (4.6)$$

For operation in the active region, it is clear that v_{GS} must always be greater than V_T and also $v_{DS} > (v_{DS} - V_T)$; the use of R_g satisfies this constraint (since i_G is near zero). It should be noted that this is a simplified circuit to show how to find the operation point and load line. In practice, for an enhanced MOSFET, it is difficult to stabilize the operation without a better biasing network and a stabilizing (feedback) resistance from S to ground.

One method of analysis, for an FET configured for a simple amplifier circuit, is a graphical one using the load-line (see Fig. 4.8). Consider the static characteristics of a typical n-channel enhancement MOSFET coupled to an input voltage of $v_{in} = 0.5 \sin \omega t$ through a capacitance whose reactance is negligible for the signal frequencies being considered (also consider that signal frequencies are well below where one would have to consider any internal capacitance effect of the FET). For a power source that provides 15 volts for V_{DD} and for an R_D of 2 kΩ, one may draw the load-line for each extreme of operation (that is, for $v_{DS} = V_{DD}$ and for $i_D = 0$, and when $v_{DS} = 0$ such that $i_D = V_{DD}/R_D$). It is also important to keep v_{GS}

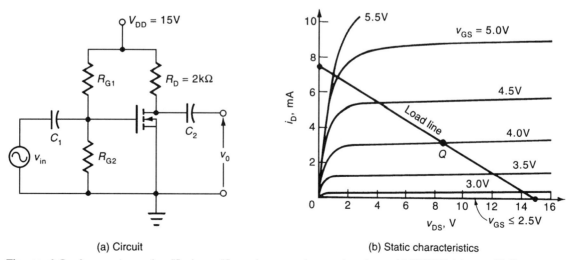

(a) Circuit

(b) Static characteristics

Figure 4.8 A one-stage simplified amplifier using an enhanced n-channel MOSFET (along with its characteristic curves)

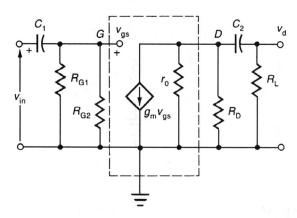

(a) Low frequency circuit model

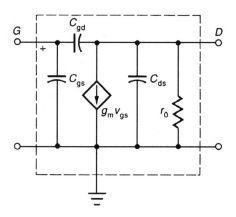

(b) High frequency MOSFET model

Figure 4.9 A small signal equivalent circuit model (for Fig. 4.8)

well above V_T so that biasing is required; one method for this biasing is to use a voltage divider network in the input, R_{G1} and R_{G2},

$$V_{GS} = [R_{G2}/(R_{G1} + R_{G2})]V_{DD} \qquad (4.7)$$

The size of these biasing resistors should be reasonably high so as not to influence the input resistance of the FET. (Actually these values may be from several hundred kilo-ohms to several meg-ohms.) Assume R_{G2} is 500 kΩ, then by Equation 4.7, R_{G1} is found to be 1.4 MΩ (actually 1.375). Then by noting the voltage change of +/– 0.5 volts of v_{gs} around the operating point, Q (for quiescent—not to be confused with the quality factor also given as Q), it is observed that v_{ds} changes from 4.4 to 12.2 volts. Because the input voltage has a peak-to-peak value of 1.0 volt, the output has a peak-to-peak change of 7.8 volts, thus a voltage gain of 7.8. Here it is obvious that the output voltage is distorted because the swing on either side of Q is not quite linear.

A small signal equivalent circuit for Figure 4.8 is shown in Figure 4.9a for lower frequencies (that is, all internal capacitances are neglected), while Figure 4.9b takes into account the internal parameters one normally considers. The resistance r_o is the resistance of the MOSFET in the active region and is approximately inversely proportional to any bias current (r_o typical varies between 10 kΩ to 100 kΩ[*]). Both the gate-to-source and gate-to-drain capacitances[*] range from less than 1 up to approximately 3 pF while the drain-to-substrate may be less than 1 pF.

The feedback capacitance, C_{gd}, is important when taking into account the Miller effect and must be considered at higher frequencies (but, of course, may still be ignored at lower frequencies). The transconductance term, g_m relates the current, i_D, to the voltage, v_{GS}. If one were to plot the static values of i_D vs v_{GS} from the load-line, g_m would be the slope (see Fig. 4.10) of the resulting curve at the operating point Q. However, in the operating region, the i_D vs v_{GS} plot may also be obtained from plotting equation 4.5; and, since g_m is the slope at point Q, the derivative of equation 4.5 yields g_m directly as

$$g_m = 2K(v_{GS} - V_T) \qquad (4.8)$$

---

[*] These typical values are from Sedra & Smith, *Microelectronic Circuits*, Holt, Rinehart and Winston, 2nd ed., 1987, pp 350–51.

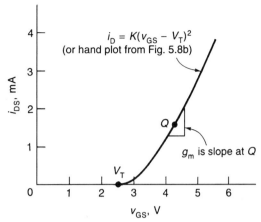

Figure 4.10   Finding g_m from a small-signal plot of Fig. 4.8b

The voltage gain from the small signal equivalent circuit is given by $|A| = v_d/v_{gs} = -g_m R_D$ if one neglects r_o; on the other hand, if r_o is included, the R_D term should be paralleled with r_o; this reduction of the new R_D, of course, tends to reduce the stage gain. For instance the slope, g_m, at the Q point in Figure 4.10 is found to be 3.75×10^{-3} S (newer notation is given in "siemens", however "mhos" is still in common use), and the resulting gain (for an R_D of 2 kΩ) gives $|A| = g_m R_D = 7.5$; the effect on R_D of r_o is negligible for this particular case.

An example for a tuned amplifier involving gain/bandwidth using a particular JFET follows (see Fig. 4.11). Assume the JFET has the following parameters: $g_m = 2.5 \times 10^{-3}$ S and the output resistance is known to be $r_o = 20$ kΩ. Also assume the pinch off voltage, v_p, is –3 volts and $I_{DSS} = 5$ mA (I_{DSS} = drain current with $G_{GS} = 0$).

Select R_s for an $I_D = 4$ mA and a V_{DD} such that $V_{DS} = 10$ V. Also, given that $R_g = 1 \times 10^6$ ohm and $L = 10 \times 10^{-6}$ H with a series coil resistance of 1.57 Ω at 5 MHz, determine the stage gain, v_o/v_s and the 3 dB bandwidth at $f_o = 5$ MHz. The solution follows.

For a JFET, $I_D \approx I_{DSS}(1 - |V_{GS}|/|V_p|)^2$

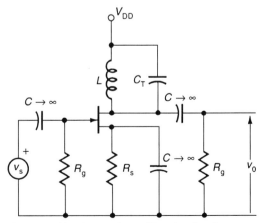

Figure 4.11   A tuned amplifier circuit using a JFET

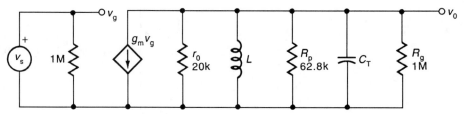

Figure 4.12 Equivalent circuit

then,

$$I_D = 4 \cong 5(1 - |V_{GS}|/3)^2$$
$$4/5 = (1 - |V_{GS}|/3)^2$$
$$0.89 - 1 = -|V_{GS}|/3, \quad |V_{GS}| = 0.32$$

then,

$$R_S = 0.32/4 \text{ mA} = 80 \text{ }\Omega \quad \text{and} \quad V_{DD} = 10 + 0.32 = 10.32 \text{ volts.}$$

For resonance,

$$\omega_o = 1/\sqrt{LC_T}, \quad \text{where} \quad \omega_o = 2\Pi(5 \times 10^6)$$
$$(10\pi \times 10^6)^2 = 1/(10 \times 10^{-6}C_T), \quad C_T = 101 \text{ pF.}$$

For the coil, the quality factor may be found from

$$Q_o = \omega_o L/R_{coil} = (2\pi \times 5 \times 10^6 \times 10 \times 10^{-6})/1.57 = 200 = Q_s$$
$$\text{For } Q > 10, \ Q_o \approx Q_s, \quad \text{or} \quad R_p/(\omega_o L) = \omega_o L/R_{coil},$$
$$R_p = Q_o \omega_o L = 200(2\pi \times 5 \times 10^6 \times 10 \times 10^{-6}) = 62.8 \text{ k}\Omega.$$

We may now draw the following equivalent circuit (see Fig. 4.12).
    The total load seen by the drain circuit is

$$R_{L\text{-effective}} = r_o \| R_p \| R_g = 1/[(1/20k) + (1/62.8k) + (1/1M)] = 14.9 \text{ k}\Omega.$$
$$v_o = -g_m v_g R_{L\text{-effective}}$$

and, since $v_g = v_s$, then $v_g/v_s = -g_m R_{L\text{-eff}} = -2.5 \times 10^{-3}(14.9k) = 37.4.$
    The effective Q of the drain circuit is

$$Q_{effective} = R_{L\text{-effective}}/\omega_o L = 14.9k/(\omega_o L) = 47.4,$$

the 3 dB bandwith is

$$\text{BW} = \omega_o/Q_{eff} \quad \Rightarrow \quad f_o/Q_{eff} = (5 \times 10^6)/47.4 = 105 \text{ kHz.}$$

Bipolar Junction Transistors (BJT)

The bipolar junction transistor, considered as a current controlled device, has several advantages over the FET type. In general, the BJTs are somewhat more linear and, in addition, they yield a higher gain and a superior frequency performance. The family of BJT's are either the npn or the pnp type (see Fig. 4.13 for the npn type) and may use several different circuit configurations' the common-emitter, the common-base, and the common-collector circuits. Most of this discussion will be for the npn type involving the common-emitter mode. Also, since

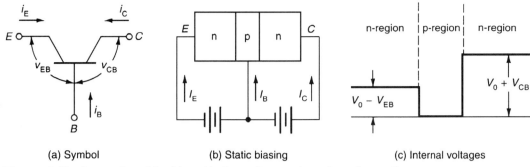

(a) Symbol   (b) Static biasing   (c) Internal voltages

Figure 4.13 An npn transistor biased in the active mode region of operation

the transistor may act as a switch in a nonlinear mode (discussed elsewhere in this book), most of this discussion will be for the linear mode of operation as used in most amplifiers.

The physical construction of the BJT is such that the base is quite narrow so that when the device is properly biased most of the injected (electron) carriers diffuse across the narrow base and into the collector area. On the other hand, the base is biased positively with respect to the emitter and holes will be injected into the emitter area; the fraction of the electrons that diffuse into the collector region while operating in the active mode is very high, usually greater than 0.98. The ratio of the collector electron current to the total emitter (electron and hole) current is given by α, the common-base current gain,

$$\alpha = I_{cn}/I_E$$

The ratio of the emitter electron current to the total emitter (electron and hole) current is given by gamma, γ, the emitter efficiency,

$$\gamma = I_{en}/I_E$$

At the same time (see Fig. 4.14), some electrons and holes, which are thermally generated near the collector-base junction, flow across the reverse biased collector-base junction; this electron-hole current will be referred to as the collector-base leakage (or reverse saturation) current, I_{CBO}. This current is very small (typically, less than 10^{-8} mA) and is temperature dependent; this current very roughly doubles for every 10°C temperature rise. For a common emitter configuration, neglecting

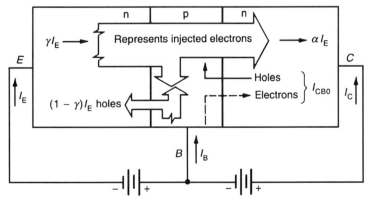

Figure 4.14 Normal operation of a properly biased npn BJT

leakage current, it is useful to define a current transfer ratio, β, (here typical values may range from less than 100 to almost 1,000) as

$$\beta = \alpha/(1 - \alpha), \text{ common-emitter gain,} \qquad (4.9a)$$

$$i_B = i_C/\beta, \qquad (4.9b)$$

$$i_E = i_C + i_B = i_C(\beta + 1)/\beta. \qquad (4.9c)$$

For a simplified current amplifier (see Fig. 4.15) and its characteristic curves, one may define an operating load-line and determine the operational quiescent point, Q, from a constant current source. For an $R_C = 2$ kΩ, a $V_{CC} = 14$ V, and a biasing current of 40 μA, one determines the quiescent point along the load-line ($I_C = 3.5$ mA, $V_{CE} = 7.3$ V) in Figure 4.15b. And with a signal current of $i_s = 20\sin \omega t\,\mu$A one may determine the current-voltage ratio for the circuit. Again, from the load-line, one may determine the change of output voltage to be 7.2 volts (10.8 − 3.6). The corresponding change of collector current is 3.5 mA (5.2 − 1.7); then as a current amplifier for a peak-to-peak current input of 40 μA and an output of 3.5 mA, the current amplification is 88, and the current-voltage ratio is 5.5 μA/V.

For the npn silicon BJT, it is interesting to note that if the collector junction is disconnected, the base-to-emitter circuit acts very nearly the same as a silicon diode with a threshold voltage drop of approximately 0.7 volt when biased in the forward direction (see Fig. 4.3a). However, from Figure 4.14, if all connections are biased properly, it is clear that once conduction starts, the majority of the current will flow through the base region and most of the current flows from the collector terminal through to the emitter terminal. For the configuration of Figure 4.15a, with the current sources replaced with a voltage source, v_s, and a series resistance, R_B, placed in the base circuit to limit the current flow, one now has a voltage amplifier. However, as long as v_{BE} is greater than the threshold voltage, i_B will be equal source voltage (minus the 0.7 v_{BE} drop), divided by R_B. As an example, if $R_B = 50$ kΩ, $v_s = 2.7 + 1.0\sin \omega t$, $R_C = 2$ kΩ, and with a $V_{CC} = 14$ V, the quiescent point will be the same as in the previous problem (see Fig. 4.15b); and, the voltage gain for the circuit may now be found. Here, i_b is given by

$$i_b = 1.0\sin \omega t/50\text{k}\Omega = 20\sin \omega t\,\mu\text{A}$$

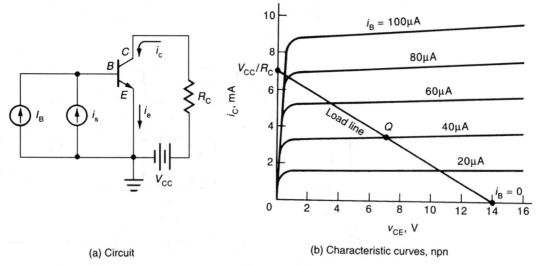

(a) Circuit   (b) Characteristic curves, npn

Figure 4.15   Simplified current-voltage amplifier using a BJT

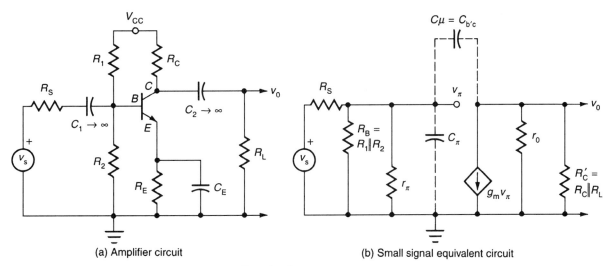

Figure 4.16 A typical BJT self biasing amplifier circuit

and it is obvious that i_b ranges between 20 and 60 μA. As before, this leads to a change of v_{ce} along the load-line of 7.2 volts (10.8 – 3.6); this results in a voltage amplification of 7.2/2 = 3.6 (negative).

Capacitance Effect

The small signal model of common emitter BJT includes the internal capacitances in active mode of operation. In the forward bias mode (EBJ) there are two equivalent parallel plate capacitances, both the depletion and the diffusion ones that may be added together, and the reverse biased (CBJ) has a depletion capacitance. The total EBJ capacitance is designated as C_π and the CBJ depletion layer capacitance is designated as C_μ (or sometimes as $C_{b'c}$). Typical values of these capacitances depend on the physical size and the operating voltages of the BJT; however, C_μ usually is 1 or 2 picofarads (small indeed, but playing an important role due to the Miller effect), while C_π has a much wider range of variation, from a few to several tens of picofarads.[*] A typical BJT common emitter voltage amplifier arrangement is shown in Figure 4.16 along with its small signal equivalent circuit.

Transconductance

The transconductance, g_m, for this BJT circuit is defined differently than for JFET. Here, the transconductance is directly proportional to the collector dc bias current, I_C, and is relatively high compared to that JFET family. The constant of proportionality is inversely influenced by the thermal voltage relationship but may be considered constant for small signal situations. Again, it is the slope of a plot of i_C vs v_{BE} curve at the Q operation point.

The internal output resistance, looking back into the collector, is given by r_0, and is approximately equal to the dc ratio of V_A/I_C. The voltage V_A is actually the $-v_{CE}$ point that represents the convergence of all of the asymptotes of the apparent straight lines of the characteristic i_C vs v_{CE} curves of Figure 4.15b. These asymptotes converge out at –50 to –100 volts (with the lower number being for pnp transistor); thus V_A is about 50 to 100 volts (positive) and is sometimes referred

---

[*]For these typical capacitance values, reference is again made to Sedra & Smith's text (page 439).

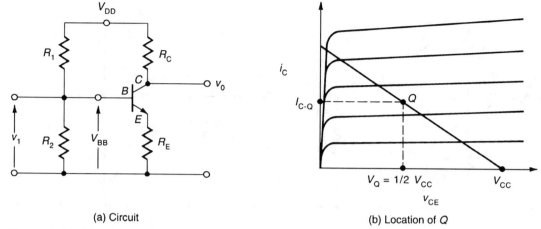

(a) Circuit   (b) Location of Q

Figure 4.17   The dc parameters of Figure 4.16a

to as the Early[*] voltage. The internal small signal input resistance, r_π, between the base and emitter (looking into the base), may be found as follows (r_e is defined in Eq. 4.10):

$$R_{in} = r_\pi = v_{be}/i_b = \beta/g_m = (1 + \beta)r_e \text{ (without } R_E) \tag{4.10a}$$

$$R_{in} = r_\pi + (1 + \beta)R_E = (1 + \beta)(r_e + R_E), \text{ (with } R_E) \tag{4.10b}$$

Before continuing the discussion on the small signal model, it is important to first determine all dc biasing values for Figure 4.16a. The resistors R_1, R_2, R_E, and R_C will play the key role for the dc values (see Fig. 4.17a). Three assumptions, or "rules of thumb," will be used.

1. For a linear amplifier (assume a common emitter configuration), one normally chooses an operating Q point at $V_{CE-Q} = (1/2)V_{CC}$ for future maximum voltage swings.

2. Select R_1 and R_2 (here, R_B will be defined as $R_B = R_1 \parallel R_2$) such that current to the base is only about 10 percent of the current through R_1 and R_2; this choice will give the base voltage good stability. Because the current through R_1 and R_2 is to be approximately 10 times greater than i_B, the relationship between R_B and R_E may be given as $R_B < 0.1 \, \beta R_E$.

3. For high values of β, $i_E = i_C$, and $i_B = i_C/\beta$, then the base voltage to ground, V_{BB}, may be found using Thevenin's voltage resistance relationship. That is, the Thevenin voltage is

$$V_{OC} = V_{BB} = [R_2/(R_1 + R_2)]V_{CC}, \tag{4.11}$$

and Thevenin's resistance is

$$R_{eq} = R_1 R_2/(R_1 + R_2) = R_B. \tag{4.12}$$

Or, solving for R_1 and R_2,

$$R_1 = (V_{CC}/V_{BB})R_B, \tag{4.13a}$$

$$R_2 = R_B V_{CC}/(V_{CC} - V_{BB}). \tag{4.13b}$$

---

[*]"Early" named for Jim Early of Bell Labs.

Because of the Thevenin equivalent series circuit, it is easily shown that (recall, $i_C = i_E$)

$$V_{BB} = R_B i_B + v_{BE} + i_E R_E = V_{BE} + (R_B/\beta + R_E)i_C. \qquad (4.14)$$

For example, assume $R_C = 2$ kΩ, $R_E = 200$ Ω, $V_{CC} = 14$ V, and $\beta = 100$; the problem is to find the values of the biasing resistors R_1 and R_2. Here, for maximum voltage swing of v_{CE}, $V_{CE-Q} = (1/2)V_{CC} = 7$ V,

$$I_{C-Q} = (V_{CC} - V_{CE})/(R_E + R_C) = (14 - 7)/(2000 + 200) = 3.18 \text{ mA}$$
$$R_B = 0.1 \ \beta R_E = 0.1 \times 100 \times 200 = 2 \text{ k}\Omega.$$

From Equation 4.14, V_{BB} may be found,

$$V_{BB} = V_{BE} + I_{C-Q}(R_B/\beta + R_E) = 0.7 + (3.18 \times 10^{-3})(2000/100 + 200) = 1.4 \text{ V}.$$

Then, from Equations 4.13a and 4.13b, $R_1 = 20$ kΩ and $R_2 = 2.22$ kΩ.

Sometimes the BJT circuit is represented by a hybrid-π model (more on this subject later); however, at this point it is important to relate certain aspects of the two models. The output current from the dependent source is $g_m v_\pi$ which is equal to $g_m r_\pi i_b$ and is also equivalent to βi_b; this leads to the equivalency of the two circuits in Figure 4.18.

Two important concepts should be pointed out when working with small signal models. First, the model does not usually show dc values; and second, when working with load-line analysis, two separate load-lines need to be considered if there is a load resistance, R_L separated from R_C with a capacitor. When determining the dc bias and the operation point Q, one superimposes a load-line for R_C on the characteristic curves; then, for an ac small signal analysis, the parallel combination of R_C and R_L is superimposed on the same characteristic curves and passes through the previously determined operation point, Q.

When working any examination problems, one should be familiar with as many different notations as possible. For instance in transistor circuits, the equivalent small signal base-emitter resistance, r_e, is sometimes specified as the equivalent diode resistance, rd, and may be given as,

$$r_e = rd = 0.026/I_E, \qquad (4.15)$$

where I_E is the dc bias value of the emitter current. Another quantity that may need to be calculated is the Miller multiplying effect on C_μ (or $C_{b'c}$); the effective value of the coupling capacitor is given as

$$C_{\text{Miller}} = C_{\text{eq}} = C_\mu(1 + gmR'_C) = C_\mu(1 + \alpha R'_C/r_e). \qquad (4.16)$$

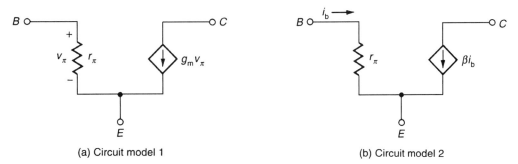

(a) Circuit model 1   (b) Circuit model 2

Figure 4.18 Two forms of an equivalent partial BJT model

Frequency Response

In general, the following remarks about frequency response apply equally well to both FET and the BJT family. Specifically, however, references will be made to the BJT of the common emitter circuit of Figure 4.16. Three frequency ranges will be considered for the single stage amplifier: the low-frequency gain, A_L, the mid-frequency gain, and the high-frequency gain, A_H. At zero frequency, one considers the biasing arrangement with all capacitors as open circuits as already presented. However, the effect of C_1, C_2, and the "grounding capacitance," C_E, is important to consider (the internal capacitances are neglected). The effect of these capacitances will be to produce poles and a zero; the gain at low frequency will be lower than the gain at mid-frequency due to the influence of these poles and zero, resulting in a lower gain that eventually reduces to zero at dc. Without the pole-zero analysis, it is obvious that the coupling capacitors C_1 and C_2 become open circuits at zero frequency.

For mid-frequency one normally treats all external capacitances such that all of their reactances are zero (*i.e.*, $C \rightarrow$ infinity) and all internal capacitances open circuits. The gain found under these constraints is the reference or mid-frequency gain, A_M.

For high frequencies, one treats all external capacitive reactances as zero (C's $\rightarrow \infty$) and only considers the internal capacitances. The Miller effect normally must be considered for the coupling capacitor C_μ (see Eq. 4.16). The total effective capacitance, C_T, will then be the sum of C_π and C_{eq}. The effect is to produce a pole, $1/(1 + s/\omega_H)$, and to multiply the mid-frequency gain by this quantity,

$$A_H = (\omega_H/(s + \omega_H)A_M \qquad (4.17)$$

Here, if one makes a series resistor, R', equivalent to all of the resistors to the left of v_π and multiplies it by C_T, the pole is given as $\omega_H = 1/(C_T R')$.

Hybrid Models

Before giving a numerical example problem, another method of solving transistor problems, especially for BJT devices (however, JFETS are not excluded), should be presented; this notational presentation involves active devices with hybrid or "h" parameter models (see Fig. 4.19).

The defining equations are

$$V_1 = h_{11} I_1 h_{12} V_2 \qquad (4.18a)$$
$$I_2 = h_{21} I_1 + h_{22} V_2 \qquad (4.18b)$$

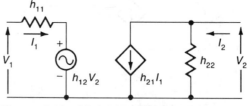

Figure 4.19 A hybrid BJT model for the active region of operation

The h_{11} parameter is an input impedance value and is usually given as $h_{ix,}$ where the x can be e, c, or b depending on which terminal of the transistor is common to input and output. Thus $h_{ie} = h_{11}$, the input impedance value for a transistor in common emitter configuration.

h_{12} is a reverse voltage feedback ration and thus h_{rb} is h_{12}, the reverse feedback ratio for a transistor in common base configuration.

The h_{21} is a forward current transfer ratio, so h_{fc} would be the forward current transfer ratio for a common collector configuration. Finally, h_{22} is an output admittance value so h_{oe} is the output admittance for a common emitter transistor while h_{ob} is for a common base, and so on.

The h-parameters are convenient to use because the gain or impedance equations are the same for all configurations. The drawback is that a deafened set of parameters must be used for each configuration. The h-parameters given are usually real quantities, although at higher frequencies they are complex.

For FET circuits, the most frequently used model to occur in these examinations, the open gate and transconductance current generator is in parallel with a drain resistance at the output. Biasing of the FET requires knowledge of the relationship of the drain saturation current, the pinch-off voltage, and the gate to source voltage.

Simplification to most electronics problems in these exams can be made by keeping in mind various impedance levels and neglecting terms that are large or small by comparison. In general, the output impedance of bi-polar transistors is reasonably high, in the order of 20k to 50k ohms for the common emitter configuration. This makes the h_{oe} term negligible when collector loads are an order of magnitude less. Similarly, the h_{re} term can usually be neglected when the collector loads are small. The elimination of the h_{oe} and h_{re} parameters reduces the low frequency model of the transistor to a very simple circuit allowing for a fast approximation of the overall gain and impedance values of an amplifier. These quick estimates are usually within 10 percent of the values obtained by using the more complete model and parameter, whose values are seldom known to 10 percent accuracy.

Sometimes the h-parameters are given in terms of common base or collector values, while a problem might be presented in a common emitter configuration. Although some caution needs to be exercised, one may need to convert from one base to another by use of any standard conversion table available in most electronic handbooks (see Table 4.1). Parameter conversion from the hybrid model to the more contemporary hybrid-π small signal model (using r_π, etc.) may be shown to be (for the common emitter configuration),

$$g_m = I_C/0.026, \ r_\pi = h_{fe}/g_m, \ r_x = h_{ie} - r_\pi,$$
$$r_\mu = r_\pi/h_{re}, \ 1/r_o = h_{oe} - h_{fe}/r_\mu = I_C/V_A.$$

For the hybrid-π model as previously discussed (see Fig. 4.18), the main difference is that the current controlled source is replaced by a voltage controlled source. Frequently, using the hybrid-π model, a quantity called the base spreading resistance, $r_{bb'}$ (same as r_x), is used in some examination problems. This $r_{bb'}$ models the base region between the base terminal and a fictitious terminal, b', at the boundary of the base-emitter junction; the $r_{bb'}$ is much smaller than r_π and may be negligible at low frequencies.

A numerical example of a broadband amplifier using the hybrid-π model of a BJT follows (see Fig. 4.20).

Table 4.1 Transistor Small Signal Characteristics
(Numerical values are for a typical transistor operating under standard conditions)

Symbols			Common Emitter	Common Base	Common Collector	T-Equivalent
h_{11e}	h_{ie}	r_i	1500 ohms	$\dfrac{h_{ib}}{1+h_{fb}}$	h_{ic}	$r_b+\dfrac{r_e}{1-\alpha}$
h_{12e}	h_{re}		3×10^{-4}	$\dfrac{h_{ib}h_{ob}}{1+h_{fb}}-h_{rb}$	$1-h_{rc}$	$\dfrac{r_e}{(1-\alpha)r_c}$
h_{21e}	h_{fe}	A_I	49	$-\dfrac{h_{fb}}{1+h_{fb}}$	$-(1+h_{fc})$	$\dfrac{\alpha}{1-\alpha}=\beta$
h_{22e}	h_{oe}	$\dfrac{1}{r_o}$	30×10^{-6} mho	$\dfrac{h_{ob}}{1+h_{fb}}$	h_{oc}	$\dfrac{1}{(1-\alpha)r_c}=\dfrac{1}{r_d}$
h_{11b}	h_{ib}	r_i	$\dfrac{h_{ie}}{1+h_{fe}}$	30 ohms	$-\dfrac{h_{ic}}{h_{fc}}$	$r_e+r_b(1-\alpha)$
h_{12b}	h_{rb}		$\dfrac{h_{ie}h_{oe}}{1+h_{fe}}-h_{re}$	5×10^{-4}	$h_{re}-1-\dfrac{h_{ic}h_{oc}}{h_{fc}}$	$\dfrac{r_b}{r_c}$
h_{21b}	h_{fb}	A_I	$-\dfrac{h_{fe}}{1+h_{fe}}$	-0.98	$-\dfrac{1+h_{fc}}{h_{fc}}$	$-\alpha$
h_{22b}	h_{ob}	$\dfrac{1}{r_o}$	$\dfrac{h_{oe}}{1+h_{fe}}$	0.5×10^{-6} mho	$\dfrac{h_{oc}}{h_{fc}}$	$\dfrac{1}{r_c}$
h_{11c}	h_{ic}	r_i	h_{ie}	$\dfrac{h_{ib}}{1+h_{fb}}$	1500 ohms	$r_b-\dfrac{r_e+R_l}{1-\alpha}=R_l(\beta+1)$
h_{12c}	h_{rc}		$1-h_{re}\approx1$	1	1	$1-\dfrac{r_e}{(1-\alpha)r_c}$
h_{21c}	h_{fc}	A_I	$-(1+h_{fe})$	$-\dfrac{1}{1+h_{fb}}$	-50	$-\dfrac{1}{1-\alpha}=-(\beta+1)$
h_{22c}	h_{oc}	$\dfrac{1}{r_o}$	h_{oe}	$\dfrac{h_{ob}}{1+h_{fb}}$	30×10^{-6} mho	$\dfrac{1}{(1-\alpha)r_c}$
	α		$\dfrac{h_{fe}}{1+h_{fe}}$	$-h_{fb}=-h_{21b}$	$\dfrac{1+h_{fc}}{h_{fc}}$	0.98
	r_c		$\dfrac{1+h_{fe}}{h_{oe}}$	$\dfrac{1-h_{rb}}{h_{ob}}=\dfrac{1}{h_{22b}}$	$-\dfrac{h_{fc}}{h_{oc}}$	1.7 MΩ
	r_e		$\dfrac{h_{re}}{h_{oe}}$	$h_{ib}-\dfrac{h_{rb}}{h_{ob}}(1+h_{fb})$	$\dfrac{1-h_{rc}}{h_{oc}}$	10 Ω
	r_b		$h_{ie}-\dfrac{h_{re}}{h_{oe}}(1+h_{fe})$	$\dfrac{h_{rb}}{h_{ob}}=\dfrac{h_{12b}}{h_{22b}}$	$h_{ic}+\dfrac{h_{fc}}{h_{oc}}(1-h_{rc})$	1 KΩ

r_i = input resistance r_c = collector resistance   power gain = $G = A_V A_I$

r_o = output resistance r_e = emitter resistance $\beta = \dfrac{\alpha}{1-\alpha}$, $\alpha = \dfrac{\beta}{\beta+1}$

A_V = voltage amplification r_b = base resistance R_l = load resistance

A_I = current amplification α = short circuit current multiplier

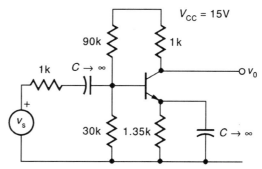

Figure 4.20 A single stage amplifier circuit

Given the following transistor parameters,

r'_{bb} = (Base spreading resistance) = 200Ω,
f_t = (short circuit current gain bandwidth product) = 100×10^6 Hz,
β = (common emitter low frequency current gain) ≈ h_{fe} = 99
C_μ = $C_{b'c}$ (depletion layer capacitance) = 5 pf (at given bias point),
V_{BE} = (diode drop forward biased base emitter junction) = 0.6 volts.

find the midband gain, v_o/v_s. at the 3 dB bandwidth (assuming that the low frequency cutoff ≈ 0 Hz) of this amplifier using the hybrid Pi model for the transistor.

Solution
One parameter needed for the hybrid Pi model that is not given explicitly is the incremental base-emitter diode resistance rd the same as r_e (refer to Equation 4.15),

$$r_e = rd \approx 0.026/I_E \ \Omega,$$

where I_E is the dc bias current.

Thus we must first calculate the correct bias point for this transistor. The equivalent circuit for this calculation is shown below. We use a Thevenin equivalent for the biasing network.

$$I_B = (V_{Th} - V_{BE})/[R_{Th} + r_{bb'} + R_E(\beta + 1)] = (3.75 - 0.6)/[22.5k$$
$$+ 200 + 100(1.35k) \ 20 \ \mu A]$$

Then,

$$I_E = (\beta + 1)I_B = 2 \text{ mA}, \quad \text{and} \quad rd = 0.026/(2 \times 10^{-3}) = 13 \ \Omega.$$

Now the equivalent hybrid-π circuit is given in Figure 4.22.

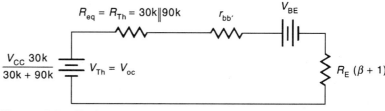

Figure 4.21 Equivalent circuit for dc bias

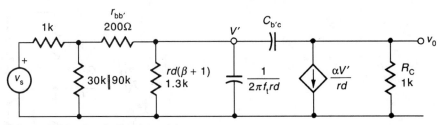

Figure 4.22   Small signal model

We may simplify this equivalent circuit by assuming that the bias network is large compared to z_{in} (22k >> 1.5 k) and assuming that R_C is small compared to the output impedance of the transistor and impedance of $C_{b'c}$ so that all of the current from the controlled source flows through R_C. This enables us to unilaterize the circuit and replace $C_{b'c}$ with its miller value as seen by the input circuit.

$$C_{miller} = (1 + \alpha R_C/rd)C_{b'c}$$

and if

$$\alpha R_C/rd >> 1, \text{ then } C_{miller} \approx (\alpha R_C/rd)C_{b'c}$$

Then the total capacitance seen at the input is

$$C_T = 1/2(2\pi f_t rd) + \alpha R_C C_{b'c}/rd,$$
$$\text{or, } C_T \approx (1 + \alpha R_C C_{b'c}\omega_t)/(\omega_t rd) = 507 \text{ pf.}$$

Now the equivalent circuit looks like Figure 4.23.
    For mid-band, $|1/(j\omega C_T)| >> rd(\beta + 1)$,

$$V' = v_s(1.3k)/(1k + 200 + 1.3k) = 0.52 \text{ } v_s,$$
$$v_o = -\alpha R_C V'/rd = -0.52v_s(1k)/13,$$

or,

$$v_o/v_s = -40.$$

The upper 3 dB point $|1/(1/(j\omega C_T)| = 1.3 \parallel 1.2k$,

or, $\omega_{3dB} = 1/\tau = 1/[(507 \times 10^{-12})(1.3 \parallel 1.2k)] = 3.16$ radians/sec $\equiv 503$ kHz.

In conclusion, any linear electronics circuits can be analyzed in stages by starting with the output stage and first determining the effective load seen by the stage. Then determine the Thevenin equivalent source impedance that drives the stage, and subsequently compute the state gain. Then, using a similar technique, work on the stages that drives the output stage and so forth.

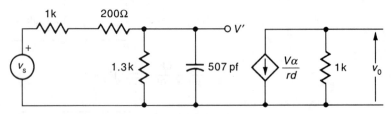

Figure 4.23   Modified small signal model

If the amplifier has feedback, calculate the open loop gain as above with the feedback disconnected, but allowing for the loading effects of the feedback network. Then determine the feedback type, voltage or current, series or shunt, and the feedback ratio. You may then predict the closed loop behavior in accordance with the classical feedback relationship

$$G = A/(1 + AF),$$

where A is the open loop gain and F is the feedback ratio.

The effect of feedback on input and output impedances is similar to that of gain modification by the factor $(1 + AF)$. Care must be used to determine whether the factor increases or decreases the quantity in question. The following rules should help.

1. If the feedback is derived from the output voltage, then the output impedance is reduced by this factor. If the feedback is derived from the output current, then the output impedance is increased by this factor.

2. With respect to the input, if the feedback voltage is in series with the input source, the impedance is increased and if it is in shunt with the input, the input impedance is lowered. The classical example of the latter is the operational amplifier used in the inverting mode. The input terminal is a virtual short circuit to ground due to the large magnitude of the loop gain. This type of amplifier is widely used and the concept will be more fully presented in the next section.

INTEGRATED CIRCUITS AND OPERATIONAL AMPLIFIERS

Integrated circuits, ICs, have more than one component on a chip. As a simple classification, a small-scale IC contains up through 60 components, while a medium-scale integrated circuit, MSI, contains up through 300, between 300 and a 1,000 would be a large scale (LSI), and over a thousand would be referred to as very large scale (VLSI) integrated circuits. These ICs may be linear and/or nonlinear and, of course, are used in almost all devices today. The kinds and applications of these devices are too broad to cover in a short review such as this; however, in this section, the operational amplifier (op-amp) will be selected for discussion.

The Operational Amplifier (op-amp)

The operational amplifier is a high gain differential amplifier circuit that has been highly developed over the years. Since the cost has gone from a few hundred dollars (old vacuum tube era) to less than a dollar for highly developed integrated circuits, the applications for this device cover almost all areas of engineering and thus may well be on the examination—especially when instrumentation type questions are presented. An op-amp has several characteristics to fit into the category of the op-amp designation. These characteristics include the following:

1. The amplification is very high and usually is in the order of 100,000 or more. This amplification is based on the input voltage being the difference between two very small voltages, $v_d = v_1 - v_2$. These voltages are designated as being

positive and negative; however, the actual polarity of the applied voltages could be either. Of course this implies that the output voltage is zero if the two small input voltages equal each other (see item 5 for exceptions).

2. The amplification is flat over a frequency range from zero Hz to some higher frequency (see gain bandwidth product definition).

3. The currents to the actual input pins (both positive and negative) are very small and may usually be neglected; this means that the input resistance (or impedance) is very high. Also, the output resistance (or impedance) is very low; again, low means relatively small as compared with the external circuitry values.

4. The device is linear over a known range. This means that the super position theorem applies over a range of voltages before saturation voltage is reached; this does not mean the op-amp is limited to linear operation.

5. The device has a common mode rejection ratio, CMRR, whose effect is small, and is frequently neglected; however, in certain applications it should be considered. The common mode rejection ratio is defined as the ratio of the differential gain to the common mode gain (may be expressed in dB's). More on this subject later.

One may visualize an op-amp as the equivalent small signal model shown in Figure 4.24a. In this figure, the positive and negative power supply (+/– PS) connections are shown, but that isn't always the case (see Fig. 4.24b). Because input voltages may be +/–, the power supply voltages are both with respect to some common point, usually referred to as signal ground.

The power supply voltages must be larger than the largest output voltage expected (for linear operation), within the limits of the op-amp. These constraints are normally assumed and are not usually part of a problem or even shown on the diagram (see Fig. 4.24b).

Linear Operation

Primarily the (linear) op-amp is used in circuits designed for use in either the inverting or the noninverting mode. Consider the inverting mode circuit (see Fig. 4.25) here, since the two voltage input pins are assumed to go to an open

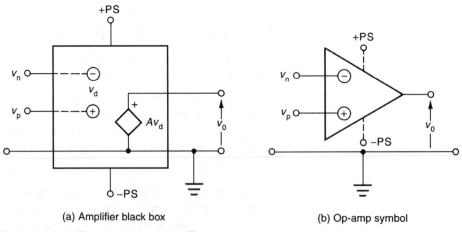

(a) Amplifier black box   (b) Op-amp symbol

Figure 4.24   The operational amplifiers

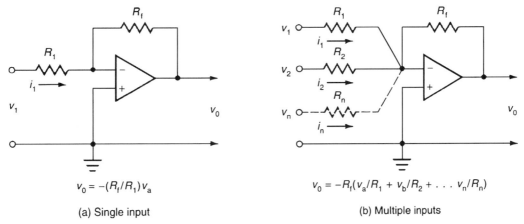

$$v_0 = -(R_f/R_1)v_a$$

(a) Single input

$$v_0 = -R_f(v_a/R_1 + v_b/R_2 + \ldots v_n/R_n)$$

(b) Multiple inputs

Figure 4.25   The op-amp in the inverting mode

circuit, the input currents for both the plus and minus pins are near zero, and the voltage difference between these two pins may only be a few micro-volt, or compared to external voltages, are essentially zero! This makes the circuit analysis especially easy in that if one knows one input voltage, the other is approximately the same. Thus, if one input happens to be grounded (as in Fig. 4.25a), the other input level is almost zero and is sometimes referred to as virtual ground. The equations for the output voltage given in Figure 4.25a are easily obtained. Because the current into the op-amp itself is extremely low or considered to be zero, then, by summing currents at node #1, i_i must equal $-i_f$ and the voltage at the junction is virtually zero volts. For this zero voltage node, i_1 is easily found to be v_1/R_1 and i_f is found to be v_o/R_f. Thus we have created a circuit that amplifies (by a factor of $-R_f/R_1$) as it inverts and has a relatively low current input at v_1 (to the external circuit if R_1 is high) and whose output acts almost like an ideal voltage source. For more than one input, the currents at node #1 are summed to yield the equation in Figure 4.25b.

As an example problem (for Fig. 4.26), assume four different transducers, each producing a possible maximum ac voltage of 0.1, 1.0, 5.0, and 10 volts respectively. Assume each transducer needs to be recorded on a one channel recorder whose desirable input signal level is one volt but with an input impedance that won't allow a direct connection to the transducers. Because only one input at a time may be recorded, all that is necessary is to determine each resistance ratio for the summing op-amp circuit. Here, summing is not required but multiplying by a constant is. The ratios are easily found. As an example, for changing

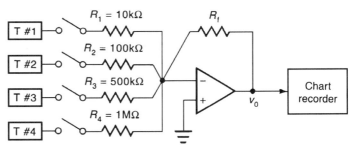

Figure 4.26   An op-amp circuit for matching voltage levels of various transducers (Caution: only one switch to be closed at any one time.)

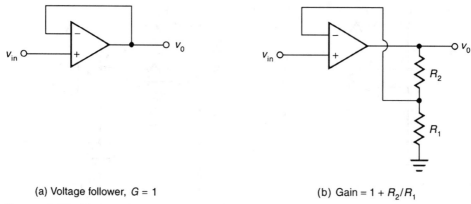

(a) Voltage follower, $G = 1$ (b) Gain $= 1 + R_2/R_1$

Figure 4.27   Noninverting op-amp circuits

the level of the first voltage from 0.1 to 1, the ratio is obviously a factor of ten. The common feedback resistor is arbitrarily chosen to be 100K ohms (the typical range of values run from a few kilo-ohms to several megohms), then R_1 must be 10K ohms. Calculating the other resistor ratios give the numerical values shown in the circuit of Figure 4.26. It should be noted that when using the summing configuration of the op-amp as a summer, all inputs connected at the same time, the dc and/or the ac, the instantaneous algebraic sum of the input voltages multiplies by the resistor ratio values will be summed together, they should not exceed the specified linear output voltage.

The other standard type circuit configuration for the op-amp is the noninverting mode. For this kind of circuit operation the negative input pin is normally fed from the output (perhaps through a resistive network as in Figure 4.27b) and the plus pin terminal is connected as an input. For the directly fed back design, the output voltage, being also the minus terminal input, must be within a few microvolts of the plus terminal. Thus, this circuit has an output essentially the same as the input; therefore, this circuit has a gain of unity with no change in polarity! The circuit is usually referred to as a buffer or voltage follower. The advantages are that it takes almost no power from an input source and the output is almost as though it were an ideal voltage source. For the circuit shown in Figure 4.27b, the minus input voltage must follow a certain percentage of the output and is considered a noninverting amplifier whose gain is given as $v_o/v_{in} =$ Gain $= 1 + R_2/R_1$. As an example, assume it is desired to have a noninverting op-amp with a gain of three; it is obvious that resistor ratio must then be two. If one assumes that R_2 is 10k ohms, then R_1 must be 5k ohms.

Nonlinear Operations

When analog computers were used for control system simulation, one often used op-amp simulation blocks to represent a saturated (or voltage-limited) amplifier-motor combination or whatever. This was often done within the linear region of operation by using voltage biased diodes. Because some form of this concept could appear in an examination problem, the following equivalent circuit is presented for analysis (see Fig. 4.28). For this circuit bias voltages are shown with batteries for clarity, but of course in the actual simulation circuit, these are replaced with potentiometers connected to reference voltage sources. The slope of the linear portion of the curve is given by the ratio of R_f/R_1; however, as soon as the output

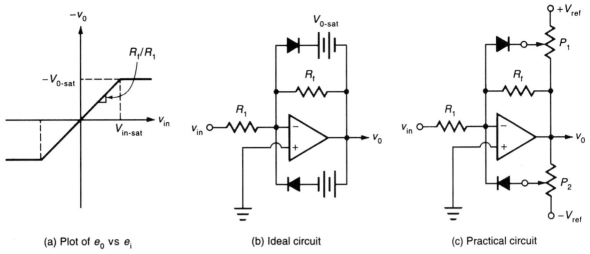

(a) Plot of e_0 vs e_i (b) Ideal circuit   (c) Practical circuit

Figure 4.28 Simulation of saturation using a linear op-amp

voltage reaches the desired saturation level (or exceeds the dc bias level), the diode conducts and effectively makes R_f appear as near zero and the gain is near zero. Of course these saturation levels may be set for different slopes, rather than zero, on either side of the axis; they may be set by inserting a resistance in series with the diode. (In practical circuits, the diode threshold voltage in the resistance of the potentiometers must be considered.)

Specialized Application

Another interesting application of the op-amp is to make a capacitor act like an inductance[*] (an impedance converter). By using two ideal op-amps in the configuration of Figure 4.29, and by assuming that the voltage difference between V_+ and v_- is essentially zero and that the currents into the op-amps are negligible, the following relationship may easily be established:

$$V_{in} = V_2 = V_4,$$
$$I_{in} = I_1, I_2 = I_3, I_4 = I_5.$$

Then it may be shown that

$$Z_{in} = V_{in}/I_{in} = (R_1R_3R_5)/(Z_2Z_4).$$

If either Z_2 or Z_4 is a capacitive reactance, say Z_2 (and the other a pure resistance), then Z_{in} is given as

$$Z_{in} = j\omega CR_1R_3R_5/R_4 = j\omega L_{eq}, \quad \text{where} \quad L_{eq} = CR_1R_3R_5/R_4.$$

In working with op-amp examination problems, specifications are frequently given. At least three of these specifications that should be understood are the interpretation of CMRR, slew rate, and gain-bandwidth product.

---

[*]This concept of designing an impedance converter is discussed more fully in a text by Savant, Roden, Carpenter, *Electronic Circuit Design*, Benjamin/Cummings Publishing Company, 1987, p. 350.

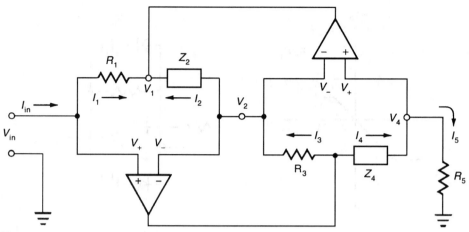

Figure 4.29 Impedance converter

Common Mode Rejection

The common mode rejection ratio is easily understood by referring to Figure 4.30. Here, it is assumed that if the + and − inputs are shorted, the output voltage would be zero even if v_{test} is not zero. However, in practice the output may not be zero. If one defines the ratio of v_o/v_{test} as the common mode gain G_{cm}, then the CMRR is given as

$$CMRR = |G|/|G_{cm}|, \tag{4.19a}$$

$$CMRR_{dB} = 20 \log CMRR. \tag{4.19b}$$

Slew Rate

Slew rate, SR, may be important when working with high frequencies and high changes of input voltage levels even when working in the linear range of the op-amp. Simply stated, it is the time delay of the output signal for a step input. The worst case (in the linear range) is the step starting from a large negative to a large positive voltage (that is, for the largest voltage just before the saturation level). The slew rate is defined as the rate of change of the output voltage (or slope) for a step input (worst case as described). Typical SR values are for an op-amp less than 1 V/μs.

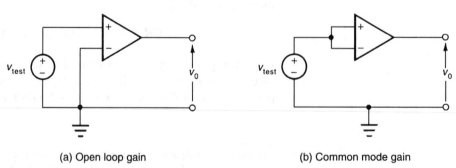

(a) Open loop gain   (b) Common mode gain

Figure 4.30 Common mode measurement

Gain-bandwidth Product

The gain-bandwidth product, GB (or sometimes GBP), is a measure of the frequency response of the op-amp. This GB is equal to the frequency at which the open loop gain, G, reduces to unity. (Actually, the interior circuitry of the op-amp contains a small capacitance to cause a roll off at some predetermined frequency to help stabilize the amplifier in open loop.) For a 741 op-amp, this -3 dB roll off point is at only 10 Hz, and continues (at -6 db/oct) until unity gain is reached at 10^6 Hz; thus, the open loop gain is given as

$$G = GB/jf = (GB/f)\angle-90° = (10^6/f)\angle-90°$$

In general, for an inverter circuit external to the op-amp with an input resistance of R_1 and feedback resistance of R_f, the cutoff frequency is given as (assuming G is still high at f_c),

$$f_c = GB/(1 + R_f/R_1). \tag{4.20}$$

The key to solving these operational amplifier circuits is to remember that the current inputs to the op-amp itself are considered to be zero. Also, that if either the plus or minus input voltage is known (or as a ratio of some other voltage) the other input voltage is considered to be the same. The application of Kirchhoff's laws to the rest of the external circuit will usually yield the correct answer.

RECOMMENDED REFERENCES

Carlson and Gisser, *Electrical Engineering Concepts and* Applications, Addison-Wesley, 1990, p. 286.

Savant, Roden, and Carpenter, *Electronic Circuit Design,* The Benjamin/Cummings Publishing Company, 1987.

Sedra and Smith, *Microelectronic Circuits*, 2nd (1987) or 3rd (1991) Edition, Holt, Rinehart and Winston.

Sedra & Smith, *Microelectronic Circuits,* Holt, Rinehart and Winston, 2nd Ed., 1987, pp. 350–51.

Control Systems

OUTLINE

The majority of this chapter is based on analysis and design methods using classical control system techniques (in contrast to the state variable method of analysis). In the past, most control system problems were classic in nature (*i.e.,* the method of solution was by root locus, Bode and Nyquist plots, etc.). However, exam problems may possibly include both state variable conversion and discrete systems involving the z-transform. For the reader who has extra study time, a very short review of state variable conversion will be presented, along with an introductory review of discrete (sampled data) systems, at the end of the chapter. No attempt will be made to cover direct digital control here. In the next chapter, some references will be made to this subject.

The background for this chapter requires a good understanding of Laplace transforms, pole-zero maps, and RLC transient response analysis. The standardized second-order differential equation solution will be the basic building block for notational purposes.

Control is concerned with regulation or control of the output, where some attribute of the output, by means of feedback, is part of this control. The analysis of a system usually involves three areas of concern:

1. *Problem Formulation*. This involves making a mathematical model of the various parameters of the system; normally one must first find the differential equation of the open loop system, then convert to Laplace transforms.

The resulting transform is then put in a flow-graph or block diagram representation (starting with the assumption that all initial conditions are zero).

2. *Determining System Stability.* Here, assuming the system is linear (and nonlinear problems are not expected on the examination), stability may be found by several methods. The three methods normally used are the Routh-Hurwitz Criteria, the root locus method of right-half pole location in the s-plane, and/or the Nyquist methods of analysis.

3. *System Performance.* One aspect of system performance is determining the amount of system error for various standard inputs. Also involved in performance analysis is percent overshoot (to a step input), time-to-the-first-peak and settling time (again, to a step input), and various other measures of performance in the time domain. Just as important is the frequency domain, which involves phase lag, magnitude ratio, resonant peak, bandwidth, and both phase and gain margin. Methods for determining performance include Bode and Nyquist plot (Nichol's charts) techniques of analysis.

PROBLEM FORMULATION

Problem formulation is usually started by calling on one's own engineering background to determine the major components needed to get a possible layout for a design concept. As an example, if a task involves positioning a dish radar antenna remotely, the major portion of the design involves selecting a drive motor, coupling the motor to the antenna drive mechanism, adding remote positioning sensors, some kind of a power amplifier, and, finally, selecting and designing some kind of controller. If the major items are the antenna (assumed already designed) and the drive motor, then the design is started by tentatively selecting a motor that is "matched" (probably through a gear train) to the antenna load. Assume one selects a dc motor, then consider the following electromechanical configuration (see Fig. 5.1).

For the electrical portion of the problem one would simply use Kirchhoff's Law as $\sum v$'s $= 0$,

$$e_{in} = v_R + v_L + e_g,$$

where e_g is the generated voltage (*i.e.*, the induced voltage within the armature) proportional to the strength of the magnetic field (here, considered to be a permanent magnet, or constant field) and armature velocity. Therefore,

$$e_g = K_g \omega = K_g \frac{d\Theta}{dt}.$$

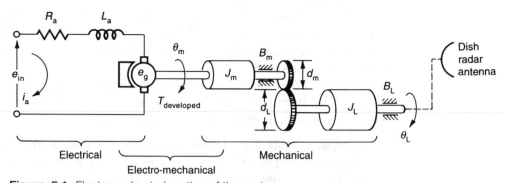

Figure 5.1 Electomechanical portion of the system

Thus,

$$e_{in} = i_a R_a + L_a \frac{di_a}{dt} + K_g \frac{d\Theta_m}{dt}. \qquad (5.1)$$

Also consider the mechanical portion of the circuit. Any torque developed by the armature is "used" to drive the load; here one simply sums torques, $\Sigma T's = 0$ (here the torque load is the antenna),

$$T_{\text{developed by armature}} = J_m \frac{d^2\Theta_m}{dt^2} + B_m \frac{d\Theta_m}{dt} + T_{\text{load}} \qquad (5.2)$$

But the load torque as seen by the motor shaft, θ_m, is reflected through the gear ratio as

$$T_L = n^2 J_L \frac{d^2\Theta_m}{dt^2} + n^2 B_L \frac{d\Theta_m}{dt}, \quad n = \frac{d_m}{d_L}$$

$$\therefore T_{\text{developed}} = (J_m + n^2 J_L) \frac{d^2\Theta_m}{dt^2} + (B_m + n^2 B_L) \frac{d\Theta_m}{dt} \qquad (5.3)$$

$$\underbrace{\qquad\qquad}_{J_{\text{effective}}} \qquad \underbrace{\qquad\qquad}_{B_{\text{effective}}}$$

Now the electromechanical relationship is such that the developed torque is proportional to magnetic flux (recall the field is considered constant) and to the armature current. Therefore,

$$T_{\text{developed}} = K_m i_a \qquad (5.4)$$

The three equations of interest are repeated below except that at this point, the manipulations become much easier if we convert to Laplace transforms. Making the simplifying assumption that all initial conditions are zero, then

$$d\theta_m/dt \rightarrow s\theta_m, \ d^2\theta_m/dt^2 \rightarrow s^2\theta_m, \ \text{etc.}$$

$$E_{in} = I_a R_a + Ls I_a + K_g s \theta_m \qquad (1)$$

$$T_{\text{developed by armature}} = J_{\text{eff}} S^2 \theta_m + B_{\text{eff}} s \theta_m \qquad (2)$$

$$T_{\text{developed}} = K_m I_a \qquad (3)$$

Combining these equations and solving for I_a, we obtain,

$$E_{in} = (R_a + Ls)I_a + K_g s \theta_m \qquad (1)$$

$$I_a = (1/K_m)(J_{\text{eff}} s^2 + B_{\text{eff}} s)\theta_m \qquad (2\&3)$$

Then,

$$E_{in} = (R_a + Ls)(1/K_m)(J_{\text{eff}} s^2 + B_{\text{eff}} s)\theta_m + K_g s \theta_m.$$

The inductive term is usually small (*i.e.,* the electric time constant is much smaller than the mechanical time constant), so if L_a is neglected, then

$$E_{in} = (R_a/K_m)(J_{\text{eff}} s^2 + B_{\text{eff}} s) \ \theta_m + K_g \ s \ \theta_m \qquad (5.5a)$$

$$= [(R_a J_{\text{eff}}/K_m)s^2 + (B_{\text{eff}} R_a/K_m)s + K_g s] \ \theta_m$$

if $K_B = (B_{\text{eff}} R_a/K_m) + K_g$.

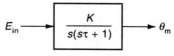

Figure 5.2 Motor-load block
diagram

Then,

$$E_{\text{in}} = [(R_a J_{\text{eff}}/K_m)s^2 + K_B s]\theta_m$$
$$= K_B\{[R_a J_{\text{eff}}/(K_m K_B)]s + 1]\}s\theta_m. \qquad \textbf{(5.5b)}$$

And, since the units of s are radians/second, then

$$\tau = R_a J_{\text{eff}}/(K_m K_B) \text{ seconds,}$$

and

$$E_{\text{in}} = K_B(\tau s + 1)s\theta_m, \qquad \textbf{(5.6a)}$$

or, the transfer function model becomes

$$\theta_m/E_{\text{in}} = (1/K_B)/[s(\tau s + 1)]. \qquad \textbf{(5.6b)}$$

Figure 5.2 shows the block diagram of our system thus far. To complete the system design, error detectors, a controller, and a power amplifier should be included for feedback control. For simplicity, a linear power amplifier and a pair of potentiometers will be considered—one connected to the output shaft, the other, to some kind of manual dial control. (However, for an actual, more sophisticated design, these devices would probably consist of error detectors being digital shaft encoders, a microprocessor as a controller, with associated A/D and D/A converters and a pulse-width amplifier used instead of a linear power amplifier. A short discussion of the digital devices is presented in the next chapter.)

If one now considers the complete system using the potentiometer error detectors and the linear power amplifier (with a variable gain control) driving the motor, the configuration might be as shown in Figure 5.3.

For the error detector, assume the two potentiometers turn through 2π and a voltage source of 6.28 volts is connected across the pot pair. Then the block diagram becomes as shown in Figure 5.4.

The system has now been defined in a block diagram form. This model may now be analyzed for system behavior directly because the system is a simple one and is called a single input, single output, SISO, system. However, for larger systems, especially those with multiple inputs and outputs, one needs to consider

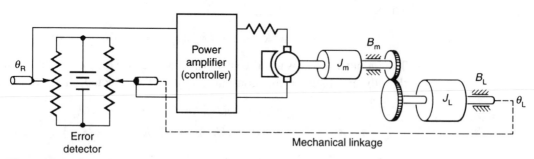

Figure 5.3 Complete simple system

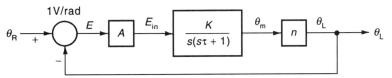

1V/rad

Figure 5.4 Complete system block diagram

block diagram or flow graph reduction techniques. (See problem 5.15 in *Electrical Engineering: Problems & Solutions.*)

The first question to be asked is, "will the system work" as configured—not necessarily how well it works; another way of asking is, "is it stable"? For a linear system, the subject of system stability analysis follows.

SYSTEM STABILITY

As previously stated, simple (linear) stability implies that if the forward transfer function has a feedback path connected around it, the system will be stable when this feedback path, H (or H may be unity), is closed. For the closed loop system of the prior problem (positioning a radar dish antenna), the concept is simple. For the closed loop transfer function, the denominator is the familiar characteristic equation. And, if this equation can be factored, and if any of the roots of the factors lie in the right-half plane (RHP) of the s-plane, the system is unstable. For the aforementioned problem, if the forward transfer function, G(s), is given as

$$G(s) = Akn/[s(\tau s + 1)], H(s) = 1.$$

Then, with reference to Figure 5.5, the closed loop transfer function is given as G_{sys} (at this point the notation(s) will be dropped for convenience).

$$G_{sys} = AKn/[s(\tau s + 1) + AKn] = K'/[s^2 + as + K'] = K'/[(s - r_1)(s - r_2)]$$

Here, the roots s_1 and s_2 are easily found from the quadratic equation and both of the poles lie in the left-hand plane (with no poles in the RHP), yielding a stable system. Of course, for most higher order systems, the characteristic equation is not so easy to factor. There are many "canned" computer programs that will factor, analyze, and optimize the system, but these will not be available during the examination. Therefore some of the classical techniques for doing these things will be discussed. One such technique for determining whether the poles of the system characteristic equation are in the LHP is the Routh-Hurwitz criterion.

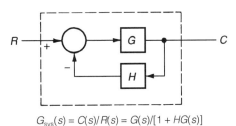

$$G_{sys}(s) = C(s)/R(s) = G(s)/[1 + HG(s)]$$
Figure 5.5 System and system equation

Routh-Hurwitz Criterion for Stability

The Routh-Hurwitz criterion gives information as to whether any poles lie in the RHP, but it does not give their actual location. Another technique, the root locus method, will give the root location in both planes, but it is far more lengthy. (It will be presented shortly.) For the Routh-Hurwitz method, one need only to set up an array from the characteristic polynomial and examine the first column of this array to determine stability. An example for a higher order system follows.

The characteristic equation (the denominator polynomial) may be factored by root locus techniques to determine if there are any roots in the right-half s-plane (which, of course, gives an unstable system). The Routh-Hurwitz method is easily implemented (rather than trying to find the factors themselves).

Consider the following simple example:

$$G = 10(s + 2)/[s(s + 1)(s + 3)], \ H = (s + 4)$$

Then,

$$G_{\text{system}} = 10(s + 2)/[s(s + 1)(s + 3) + 10(s + 2)(s + 4)]$$
$$= s(s + 1)(s + 3)/[s^3 + 14s^2 + 63s + 80].$$

The characteristic polynomial, $s^3 + 14s^2 + 63s + 80$, may be arranged in a Routh-Hurwitz array as follows (for details refer to any text on classical control):

s^3	1	63	
s^2	14	80	$x_1 = [(14)(63) - (1)(80)]/14 = 57.3$
s^1	x_1		
s^0	y_1		$y_1 = [(x_1)(80) - (14)(0)]/x_1 = 80$

Because the first column is all positive, the system will be stable in closed loop. (Try this problem for H being unity and G being three times larger—you will get an unstable system!) Of course stability may be found by frequency response techniques (*i.e.,* Nyquist criterion, etc.); but, if only the question of stability is to be answered, then use the easy *R-H* method.

Root Locus Method for Stability

The root locus method has far more functions than just finding system stability, but the method will only be briefly outlined for concept as it is applied to stability at this point. The idea is to realize that the characteristic equation (the denominator of the closed loop system function) is given as $1 + GH$ in its primitive form. This means that for values of s that make $GH = -1$, then obviously $1 + GH = 0$ and the system function goes to infinity (which defines the poles of the system). If the K (or some other parameter) in the GH function is allowed to vary, it is possible to map all locations in the s-plane (where K is real). The resulting root locus then is nothing more than the locus of all paths in the s-plane for various values of K. For any value of K that produces pole locations in the right half plane, that pole will lead to system instability, as with the R-H method. The important thing to realize is that one may determine closed loop stability from only open loop transfer functions! This concept will be more fully explored in the following section on system performance.

Frequency Response Methods for Stability

Frequency response techniques for determining system stability may be found from Bode plots, Nyquist plots, and a combined one, called a Nichol's chart plot. Using these various plots involves finding closed loop stability from only open loop equations. Like the root locus plots, much more information than stability may be obtained from these frequency domain plots. Again, the crucial point is that the denominator equation in its primitive form is $1 + GH$, leading to a mapping of a singularity function about the -1 point. It will be shown in the section on system performance for frequency response, that if the plot of GH (as the frequency varies from zero to infinity) encloses the -1 point, the Nyquist plot, the closed loop system will be unstable (except for certain unlikely exceptions).

SYSTEM PERFORMANCE

In the previous sections on system stability, the question of whether the system would be stable when the loop was closed has been answered. The system performance, assuming the system turned out to be stable, was ignored. The goal, of course, is to determine how well the system will respond to several different kinds of standard inputs (this becomes part of the system specification). Furthermore, if the system, under a number of different conditions, does not perform as desired, several modifications may be made to revise the system to meet the performance specifications (which is referred to as system compensation).

There are a number of measures of system performance to be considered. As previously mentioned, some of these measures (of the system output response for a step input) are the percent overshoot, %OS, the time-to-the-first-peak, t_p, and/or rise-time, t_r, and the settling time, T_s, of the output (see Fig. 5.6).

Frequently, when formulating the system specifications, it is convenient to have a reference standard. The nearest thing to a so-called standard has been developed for a second-order system (actually) the one shown in Fig. 5.6 and 5.12; here, %OS, t_p, T_s are all related to a quantity called zeta, ζ. The response to a second order system may be underdamped (as shown) with ζ less than one, or may

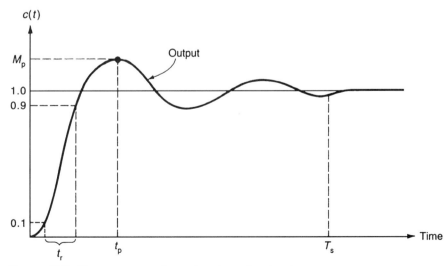

Figure 5.6 The response of a system to a step input

be critically damped with $\zeta = 1$, or overdamped with ζ greater than one (more on this subject later).

The next question to be asked after that of system stability is, "how much error will the system have to the various standard inputs" (assuming the system turns out to be stable)? The answer to this system error question follows.

System Error

System error is easily found for an ideal SISO system with unity feedback; it is merely the difference between the input and output over time. Error varies as a function of time for various standard inputs; this is called dynamic error. Dynamic error coefficients may be found for a particular system (the equations for finding these dynamic error coefficients are presented in almost any text on control systems). On the other hand, system error *after a long period of time* is easy to determine and becomes part of the specifications for a system. Again, assuming a stable system, the error, e, may be found after a long period of time by letting $t \rightarrow \infty$ and inserting this into the error time equation, but this is a lengthy procedure. However, from the final value theorem from Laplace transform theory, one may obtain the final value of e (as $t \rightarrow \infty$) directly. This final value theorem states that, for a stable system,

$$e(t) = r(t) - c(t) = \lim_{s \to 0} sE(s) \text{ for } t \rightarrow \infty, \tag{5.7}$$

where $E(s) = R(s) - C(s)$ as in Figure 5.7a. As a consequence of this equation, calculations are greatly reduced, as one does not have to take the inverse Laplace-transforms. As a short example, for the radar antenna positioning system, the error is found for a given step input of height r (here, $R(s) = r/s$) as follows (again, for convenience, the functional notation "(s)" will be dropped):

$$E = R - C = \left(\frac{1}{1+G}\right) R = \frac{\left(\frac{r}{s}\right)}{1 + \frac{AK}{[s(\tau s + 1)]}}$$

$$e(t \rightarrow \infty) = \lim_{s \to 0} s \frac{\left(\frac{r}{s}\right)}{1 + \frac{AK}{[s(\tau s + 1)]}}$$

$$= \frac{r}{1 + \lim\limits_{s \to 0} s \frac{AK}{[s(\tau s + 1)]}} \rightarrow \frac{r}{1 + \infty} \rightarrow 0.$$

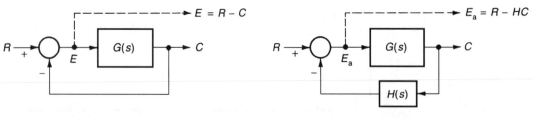

(a) E for unity feedback   (b) E_a for nonunity feedback

Figure 5.7 System error and system actuating error

The error for a constant velocity (ramp) input of $w_r t$ (here, $R = \omega_r/s^2$) is

$$E = R - C = \left(\frac{1}{1+G}\right)R = \frac{\left(\frac{\omega_r}{s^2}\right)}{1 + \frac{AK}{[s(\tau s + 1)]}}$$

$$e(t \to \infty) = \lim_{s \to 0} s \frac{\left(\frac{\omega_r}{s^2}\right)}{1 + \frac{AK}{[s(\tau s + 1)]}}$$

$$= \lim_{s \to 0} \frac{\omega_r}{s \frac{AK}{[s(\tau s + 1)]}} \to \frac{\omega_r}{0 + AK} \to \frac{\omega_r}{(AK)}.$$

And the error for a parabola, a constant acceleration, $a_r/2t^2$, (here, $R = a_r/s^3$) is

$$E = R - C = \left(\frac{1}{1+G}\right)R = \frac{\left(\frac{a_r}{s^3}\right)}{1 + \frac{AK}{[s(\tau s + 1)]}}$$

$$e(t \to \infty) = \lim_{s \to 0} s \frac{\left(\frac{a_r}{s^3}\right)}{1 + \frac{AK}{[s(\tau s + 1)]}}$$

$$= \lim_{s \to 0} \frac{a_r}{s^2 + \frac{sAK}{(\tau s + 1)}} = \frac{a_r}{0 + 0} \to \infty.$$

For this simple forward transfer function, the denominator of G has a factor "s" (to the first power) that acts as a pure integrator to any signal as it passes through G; again, for a unity feedback system, the forward transfer function is said to yield a "type 1" system classification. In short, a type 1 function yields a system with zero error for a step input, a finite error for a ramp input, and an infinite error for an acceleration input. It is easily derived that for a forward transfer function with a pure s^2 in the denominator, the error will be zero for the step and ramp input and finite for the acceleration input. A general classification is presented in Table 5.1.

It would appear that the higher the system type, the better, since more of the errors to the various inputs become zero. But this is misleading! As the power, n, of the pure number of integrations, s^n, increases (in the denominator), the system will become more unstable (resulting in closed loop poles being in the RHP).

There are short cuts for finding the finite error for the various kinds of inputs and for the different "n" type functions. However these short cuts will not be presented here. Unless one is already familiar with error analysis, the previous

Table 5.1 General classification of system error

	System Error		
Type	Step	Ramp	Accel
0	Finite	∞	∞
1	0	Finite	∞
2	0	0	Finite

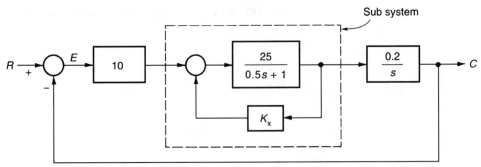

Figure 5.8 System block diagram

procedure is recommended. It allows one to calculate most finite errors that might appear on an examination. There is one caution that should be pointed out; that is, if the system does not have unity feedback, the definition of error needs to be carefully considered before attempting to work a problem. If an error is defined as the difference between the input and output, the equation for error is still $E = R - C$ (but C is no longer EG); sometimes, however, the desired error is the "actuating error" (see Fig. 5.7b), E_a, here $E_a = R - HC$. In either case one needs to modify the above final value equations to fit a particular problem. An example problem follows.

Consider the block diagram for the system in Figure 5.8. Assume it is desired to find the system final value error for the various standard inputs (a unit step, ramp, and acceleration) in terms of K_x. There are several ways to solve the problem. One is to reduce the center subsystem to its equivalent transform by block diagram or signal flow graph reduction techniques (see Fig. 5.9). Then, of course, one applies the final value theorem to obtain the answer(s) directly. The forward transfer function becomes

$$G_{\mathrm{fwd}} = 50/[(0.5s + 1 + 2K_x)s].$$

From inspection, because of the factor "s^1" in the denominator, this is a type 1 system and the error to a step is zero, the error to a unit ramp is finite, and the error to an acceleration is infinity. The finite value for a unit ramp input is then

$$e(t \to \infty) = \lim_{s \to \infty} s \left[\frac{\left(\frac{1}{s^2}\right)}{1 + G_{\mathrm{fwd}}} \right] = \left[\frac{1}{s + sG_{\mathrm{fwd}}} \right] \to \frac{1 + 25K_x}{50}.$$

Because the error is a function of K_x, in this particular case, K_x would want to be minimized. On the other hand, if the 10 and/or 0.2 in the numerator were the

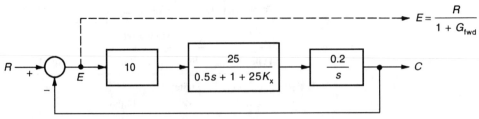

Figure 5.9 Reduced block diagram of Figure 5.8

variable, one would want to maximize these quantities for minimal error. Generally the designer has flexibility in choosing various parameters, and sensitivity to this parameter variation becomes important, as will be discussed.

Sensitivity to Parameter Variation

All of the parameters in a system may not be exactly known. Therefore it is important to know how sensitive the system is to these variations. The sensitivity, S, is defined as the ratio of the percentage change of the transfer function of the overall system to the percentage change of the particular transfer function of interest,

$$S = \frac{\frac{\Delta G_{sys}}{G_{sys}}}{\frac{\Delta G}{G}} \rightarrow \frac{\frac{\partial G_{sys}}{G_{sys}}}{\frac{\partial G}{G}} \tag{5.8}$$

For instance, if one is interested in knowing the effect on the system for a change in G, the sensitivity is found as follows:

$$S = \frac{\partial G_{sys}}{\partial G} \cdot \frac{G}{G_{sys}} = \left[\frac{1}{(1+GH)^2} \right] \cdot \frac{G}{\frac{G}{(1+GH)}} = \frac{1}{1+GH}. \tag{5.9a}$$

From this result it is clear that to reduce the effect of any change in the system transfer function one should have a large value of GH. In terms of frequency response, then, GH should be as large as possible at the frequency of interest. On the other hand, for a change in H, the sensitivity is especially important, because

$$S = \frac{\partial G_{sys}}{\partial H} \cdot \frac{H}{G_{sys}} = \left[\frac{G}{1+GH} \right]^2 \cdot \frac{-H}{\frac{G}{(1+GH)}} = \frac{-GH}{1+GH} \tag{5.9b}$$

Here, for a large value of GH, the sensitivity approaches unity, therefore any change in H is directly proportional to the change in the system. Thus any components used in the feedback path need to be as stable as possible. As an example, consider the block diagram of Figure 5.10. Assume it is desired to know the sensitivity of the system for any variations of G_1 or of H_1 at a frequency of one rad/second.

For G_1,

$$S_{G_1} = \frac{G_1}{G_{sys}} \cdot \frac{\partial G_{sys}}{\partial G_1} = \frac{1+G_3H_2}{1+G_3H_2+G_1G_2G_3H_1} = \frac{10s^3+111s^2+11s}{10s^3+111s^2+21s+10}$$

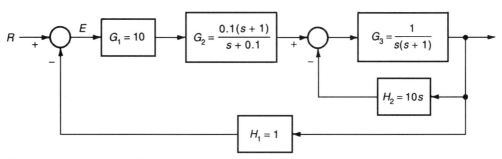

Figure 5.10 A system block diagram

at one rad/second, $|S|\,|_{s=j\omega=j1} = 1.06.$ for H_1,

$$S_{M_1 G} = \frac{H_1}{G_{sys}} \cdot \frac{\partial G_{sys}}{\partial H_1} \cdot = \frac{-G_1 G_2 G_3 H_1}{1 + G_3 H_2 + G_1 G_2 G_3 H_1} = \frac{-10(1+s)}{10s^3 + 111s^2 + 21s + 10}$$

at one rad/second, $|S|\,|_{s=j\omega=j1} = 0.136.$

In this particular case, the effect of a change in G_1 is much greater than changes in H_1 for this low frequency.

Root Locus Analysis

The expected performance of a system is related to the location of the poles and zeros of the system equation. One of the most powerful methods used in analyzing or locating these roots is finding all possible locations of the path of the roots as some parameter is varied. Frequently the gain, K, is varied for the portion of the system that is known, resulting in the root locus. However, before reviewing the root locus theory, it is first important to review the solution to an ideal second-order system for notational purposes.

Standardized curves are well known and are published for normalized second-order systems; these relationships and notation will simplify problem solutions. As previously noted, the parameter zeta relates to whether the system is underdamped ($\zeta < 1$), critically damped ($\zeta = 1$), or overdamped ($\zeta > 1$). Furthermore, by defining the undamped natural frequency as ω_n (the frequency of a pure sinusoid), one may normalize a time axis with this value and generate the standardized curves for the ideal second-order system. For a simple system (see Fig. 5.11a) and its system transfer function (see Fig. 5.11b) along with its equivalent standard notation, one may immediately identify, from these standardized curves (see Fig. 5.12), some of the system performance specifications. Furthermore, the curves are given for the time solution (eliminating the need for finding any inverse Laplace transforms). As zeta approaches zero, the undamped curve becomes sinusoidal and the time-to-the-first peak obviously is π on the normalized axis. Clearly, for damped sinusoids, the peak will always be a number somewhat larger than 3.14. If one remembers this relationship, it will frequently give a check for an approximate answer on the examination! As a short example, assume for Figure 5.11a, $G_{fwd} = 25/[s(s + 2)]$, then the system transfer function is

$$G_{sys} = 25/(s^2 + 2s + 25) = \omega_n^2 / \left(s^2 + 2\xi\omega_n s + \omega_n^2\right).$$

Here $\omega_n = 5$, $\zeta = 0.2$, from the normalized curves, one immediately determines that the percent overshoot is approximately 52 percent and the peak value occurs

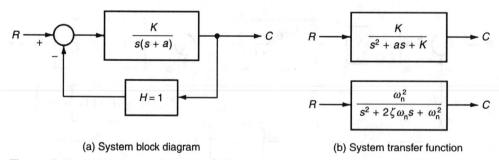

(a) System block diagram   (b) System transfer function

Figure 5.11   An ideal second-order system

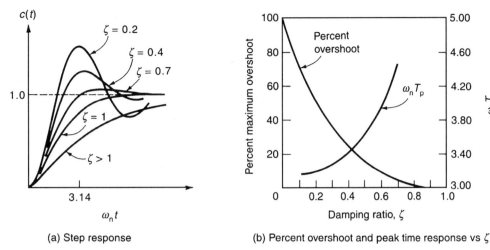

(a) Step response

(b) Percent overshoot and peak time response vs ζ

Figure 5.12 Standardized second-order curves

about 3.2 on the normalized time scale. The 3.2 value is equal to $\omega_n t$, therefore $t_p = 3.2/\omega_n = 3.2/5 = 0.64$ seconds. The settling time depends on how close to the final value $c(t)$ is acceptable. Most engineering applications require 1 or 2 percent accuracy; if the requirement is less than 1 percent, then 5 time constants away from start is used, if less than 2 percent, then 4 time constants is used. The settling time (for 2 percent) accuracy is then given as $T_s = 4/\zeta\omega_n = 4/(0.2 \times 5) = 4$ seconds. By factoring the system equation (for $\zeta < 1$), one may obtain the pole location in the s-plane.

$$G_{\text{sys}} = \omega_n^2/\left(s^2 + 2\zeta\omega_n s + \omega_n^2\right) = \omega_n^2/[s + \zeta\omega_n + j\omega_n\sqrt{1-\zeta^2})(s + \zeta\omega_n - j\omega_n\sqrt{1-\zeta^2}.$$
$$= 25/[(s+2+j4.9)(s+2-j4.9).$$

One now has the graphic relationship for both the time and the pole plot. This relationship (see Fig. 5.13) should be thoroughly studied before proceeding with the root locus review. For Figure 5.13b, from simple trigonometry, the angle of zeta is easily shown to be the $\cos^{-1}\zeta$, the reciprocal of the time constant of the system is the distance $\zeta\omega_n$ from the axis to the real axis component of the poles,

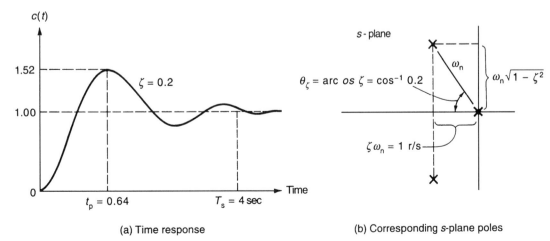

(a) Time response

(b) Corresponding s-plane poles

Figure 5.13 The second-order time and s-plane relationship

the radial distance from the axis to one of the complex poles is ω_n, and the vertical component to either of the poles is the actual damped frequency (sometimes called the "ringing frequency"), $\omega_n \sqrt{1 - \zeta^2}$. As will be shown later, a higher order problem will frequently have its complex poles (the dominant ones) related to the standardized second-ordered poles.

Again, finding the locus of the roots as some parameter is varied involves factoring. However, the root locus technique is a very clever way of finding the path of the closed loop roots from an open function without the tedious problem of actually factoring. In this review, only certain concepts will be presented to indicate the trend of the plot of the locus (although there are a number of rules used to help find the actual plot—these rules are not given as they are readily available in almost any book on control system theory). Consider a system with unity feedback. The denominator of the system transfer function is $1 + G(s)$, then, as previously discussed, the poles of the closed loop are located at values of s to make G_{sys} go to infinity, these are the same values of s to make $1 + G(s) = 0$! Because s is a complex quantity, if the absolute volume of $G(s)$ is 1 and the angle of $G(s)$ is $+/- n180°$ (where n is any odd integer), the requirement of $1 + G(s)$ being zero is satisfied. For an open loop transfer function of $G(s) = K/[s(s + a)]$, and if K may be varied from $0 \to \infty$, it is obvious the closed loop pole location(s), for an initial value $K \to 0$, will be the same as the open loop location. This is so because

$$G(s)_{\text{sys}} = G(s)/[1 + G(s)] = K/[s(s + a) + K]|_{K \to 0} \to K/[s(s + 1)].$$

This equation, of course, is trivial, as no signal could get through if K were actually zero. From this simple relationship, one may conclude that the closed loop pole path originates from the open loop poles (for $K \to 0$) and seeks their zeros (either in the finite s-plane for finite zeros, or at infinity if there are no finite zeros) as $K \to \infty$. For the previous transfer function, as K is allowed to increase, the $\angle G(s) = -n180$ relationship allows one to plot the locus with only a ruler and protractor. At this juncture, the plot apparently may be located any place in the s-plane by a trial and error procedure. It would seem that one would pick any location and test the point to see if all vectors from the open loop poles add to $+/- 180°$ (see Fig. 5.14). For Figure 5.14a, the "random" test point chosen (the author just "happened" to choose this point) has an angle $\Theta_1 = 135°$, and $\Theta_2 = 45°$ and angles

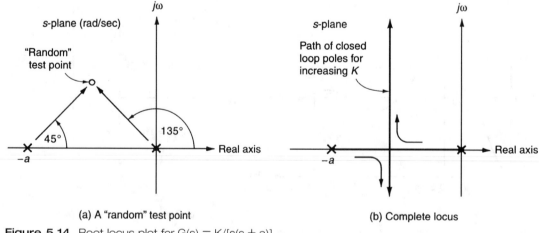

(a) A "random" test point   (b) Complete locus

Figure 5.14   Root locus plot for $G(s) = K/[s(s + a)]$

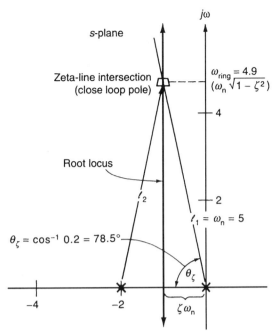

Figure 5.15 The root locus plot of G(s) = K/[s(s + 2)] with $\zeta = 0.2$

of the vectors add to the correct 180° value. For Figure 5.14b, by choosing an infinite number of test points, the complete locus may be defined. (This is not practical, but the knowledge of the rules allows one to intelligently select the "random" test points.) Suppose one wanted the system to have a certain zeta, say of 0.2. One merely finds the intersection of the zeta line ($\Theta = \cos^{-1}0.2 = 78.5°$) and the root locus to locate the closed loop pole. This value of zeta, in turn, allows one to state the %OS, the t_p, and the T_s as part of the system specifications.

For example, if $G(s) = K/[s(s + 2)]$ and for a known zeta of 0.2, the root locus (see Fig. 5.15) yields an $\zeta\omega_n$ of 1, a "ringing frequency' of 4.9 r/s, and a K yet to be found. Recall that if the absolute value of $|G(s)| = 1$, this gives $K = |s(s + 2)| = |s||(s + 1)|$ for the particular location of s. These are absolute values that represent the line lengths from the open loop poles to the closed loop pole being considered (here, $l_1 = 5$ and $l_2 = 5$). This relationship gives the value of K along the locus and is found by satisfying $|G(s)|= 1$ (then $K = l_1 x l_2 = 25$). When using the root locus, one must be careful of the form of the transfer function and use the correct K (or sometimes called K_{K-L} to emphasize the correct relationship). The following example will demonstrate the correct format.

For a unity feedback control system as shown in Figure 5.16, the form of $G(s)$ should be revised such that K (or K_{R-L}) is easily calculated from the root locus.

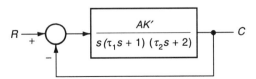

Figure 5.16 A transfer function for a third order problem

The system closed loop roots are found from solving the characteristic polynomial equation for various values of AK (but one must recall that the root locus gain "K_{R-L}" is found by writing the open loop equation in terms of discrete pole values).

$$G_{\text{open loop}} = \frac{[AK'/(\tau_1\tau_2)]}{[s(s+1/\tau_1)(s+\tau_2)]} = \frac{K_{R-L}}{[s(s+s_1)(s+s_2)]}$$

Thus, G_{system} is found as

$$G_{\text{system}} = \frac{G}{(1+G)} = \frac{K_{R-L}}{[s(s+s_1)(s+s_2)+K_{R-L}]}$$

$$= \frac{K_{R-L}}{[s^3+a_2s^2+a_1s+a_o]}$$

$$= \frac{K_{R-L}}{[(s+\alpha)(s+\beta)(s+\gamma)]},$$

where two of the roots may be either real or complex. If one plots the root locus, the loci emanates from the open loop poles to give the closed loop locus as Figure 5.17 (for brevity, the subscript "$R-L$" will be dropped).

From Figure 5.17 it is clear the root locus will yield three closed loop poles. One must always be real (to the left of $-s_2$) and the other two may be real or complex. One almost always wants the gain to be as large as feasible (resulting in a smaller error, and usually a faster responding system). Closed loop pole locations are best selected in the complex plane with a reasonably high ω_n. From experience, a good location may be found by relating the third order system to the ideal second order one by choosing an equivalent zeta line and proceeding from there to find the necessary K and the location of the other root. An approximate set of specifications derived as though it were a second order system may be found. However, how approximate the results are depends on the actual location of the third real root. If the real component of the complex roots is small compared to the third root, then the approximate specifications may well be within desired

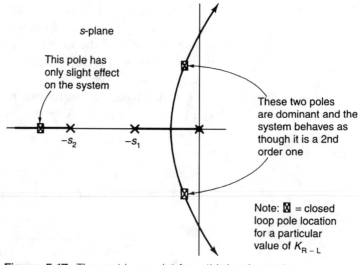

Figure 5.17 The root locus plot for a third order system

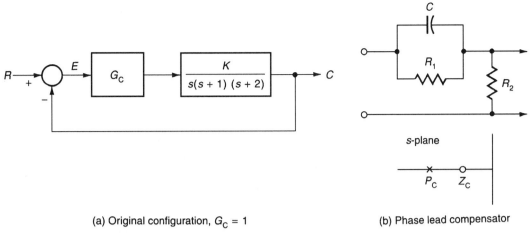

(a) Original configuration, $G_C = 1$ (b) Phase lead compensator

Figure 5.18 System configuration for the example problem

accuracy. If not, one may find the inverse Laplace transform of the three pole system and obtain the specifications from the resulting plot. A numerical example (that may need compensation) follows.

Assume the following configuration (see Fig. 5.18a) for a system that has already been selected, and which should have an overshoot to a step input not greater than 25 percent (corresponding to a zeta of 0.4) as part of its specifications. To start the analysis, the closed loop poles are located at the intersection of a zeta line of $\Theta = \cos^{-1} = 66.4°$ and the crossing of the root locus. This tentative new closed loop pole location is then evaluated for typical system performance (see Fig. 5.19) where the results (as though it were a second order system) are found to be: $K = K_{RL} = 1.35$, $t_p = 4.5$ seconds, $T_s = 13.3$ seconds, and the system error to a unit ramp input is

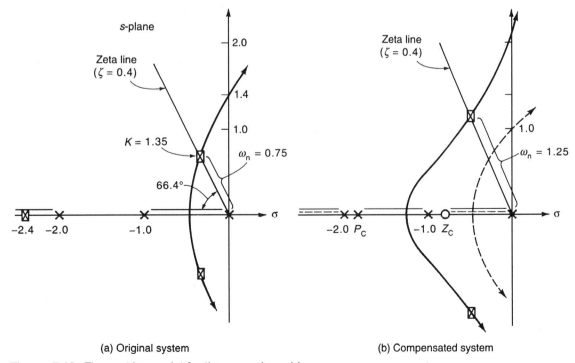

(a) Original system   (b) Compensated system

Figure 5.19 The root locus plot for the example problem

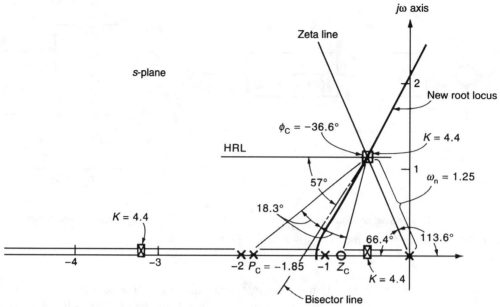

Figure 5.20   Root locus modification

1.5 radians. Assume these resulting specifications are not satisfactory for a particular additional requirement that a shorter t_p is necessary. Therefore (since the original poles are assumed fixed), some kind of modification is needed.

One such modification, or compensation, for a system is that of adding a phase-lead compensator as shown in Figure 5.18b. For this particular compensator, zero will be located between its pole and the origin of the s-plane. The "art" of selecting the $R-C$ values for pole-zero placement is not exact, but a few guidelines (depending on which parameter needs to be optimized) are in order. To increase the speed of response (*i.e.,* increasing ω_n is also reducing t_p) to a step input, is to balloon out (to the left of the original locus) the locus path (see Fig. 5.19b) for an increase in ω_n. Here, assume the desired t_p to be less than 2.75 seconds, which corresponds to a new ω_n of $1.25 r/s$. The concept of sweeping the locus away from the $j\Omega$ axis will tend to make the system more stable and will give a longer radial line (which is ω_n) from the s-plane origin to a wider root locus path. The phase-lead compensator will accomplish this action (while a phase-lag compensator has the opposite effect). The equation for this circuit (see Eq. 5.10) is as follows:

$$G_c = [s + 1/(\alpha\tau)]/(s + 1/\tau) = (s + 1.1)/(s + 4.4) \qquad (5.10)$$

where $\alpha = (R_1 + R_2)/R_2$ and $\tau = [R_1R_2/(R_1 + R_2)]C$.

One relatively easy technique (see Fig.5.20 for details) for selecting values of the compensator pole, P_C, and zero, Z_C, is given by the following "cookbook" procedure.[*] This particular outline is for increasing ω_n, but is not restricted to just that parameter.

1. First, locate the desired zeta-line on the original root locus; the radial line represents (from the origin) the closed loop ω_n. Decide how much longer this line should be for an acceptable ω_n.

---

[*] It is the author's belief that Professor Otto J. Smith (of the University of California at Berkeley) first introduced this compensation technique in the early 1960s.

2. This new desired ω_n location is where a new modified locus should pass through (recall, all angles should add to $+/-n180°$ at this new point after compensation is completed). Measure all angles from the original pole-zero plot; the sum of these angles will not equal $+/-n180°$, but the difference between the measured sum of angles and $+/-n180°$ will be the angle, Φ_C, needed for correction by the P_C and Z_C arrangement. This Φ_C should not exceed more than approximately 60° for a single phase-lead compensator. [For this example, $\Phi_C = -(113.6 + 66 + 37) + 180 = 36.6°$.]

3. At this new location (on the zeta line), draw a horizontal reference line (HRL) through this point.

4. Bisect the angle between the HRL and the zeta line. [Bisector angle is 113.6°.]

5. From this intersection and the new bisector line, draw two new radial lines at an angle of half Φ_C on either side of the bisector line. [$\Phi_C/2 = 18.3°$.]

6. At the intersection of these two half angle lines and the real axis, locate the Z_C for the line closest to the origin and P_C on the other intersection (just the reverse for a phase-lag compensator). [$Z_C = -0.8$, $P_C = -1.85$.]

The compensator $P - Z$ location should be correct; however, as a check, the sum of $P - Z$ to the new location should be checked for the 180° requirement. The values of the resistors and capacitor may now be calculated; note that one value of the R's needs to be assumed (R_2 might be a value much lower than the input resistance to the next stage). The new closed poles sites are then evaluated for typical system performance where the results (again, as though it were a second order system) are found to be $K = 4.4$, $t_p = 2.72$, $T_s = 8$ seconds, and the error = 1.0 radian. This method yields good but not necessarily optimal results. A slightly modified technique may be used to change to different zeta-line angles and different K's.

In addition to phase-lead/lag compensators, there a number of other methods commonly used. Some of these methods involve proportional, derivative, and integral controllers (use of all three are called PID controllers), feedback path compensation, active (op-amp) controllers, etc. Many other methods based on state variable analysis and optimal control theory are used (but are beyond the scope of this review). For each of the particular methods used, there are guidelines for maximizing the effectiveness of parameters selection. Of course, in practice, one would use a computer simulation program for finding the most effective compensation from examining a plot of results.

Before leaving the subject of root locus, it should be noted that for nonunity feedback with $H(s)$ being either a constant or a variable, the procedure is essentially the same as for $G(s)$ except that $H(s)G(s)$ is plotted instead. This author recommends the following manipulation for "ease of bookkeeping" for the transfer function(s):

$$G_{sys} = G/(1 + HG) = (1/H)[HG/(1 + HG)]. \qquad (5.11)$$

Here, instead of finding the root locus for G, one would plot the locus for HG and proceed as usual. Then, when all work is done, merely divide by H to obtain the system transfer function.

Steady State Frequency Analysis—Bode

Designing or analyzing a system by sinusoids (in steady state) preceded the root locus method by a number of years and is well developed. To obtain performance

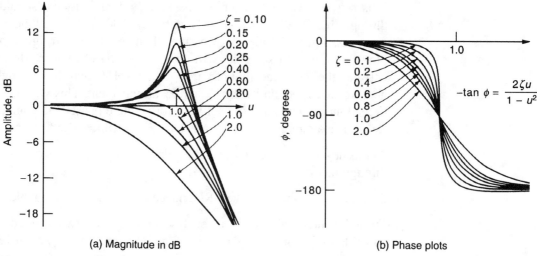

(a) Magnitude in dB

(b) Phase plots

Figure 5.21 Standardized second order frequency

specifications or when designing a system, one should have a good understanding of Bode plots and the meaning of Nyquist theory. While working in the time domain, %OS, t_p, T_s, and t_r are important specifications; in the frequency domain, the gain margin, phase margin, bandwidth, and resonant peak(s) are part of the system specifications. As an example, for the family of step response characteristics (for a second-order system) as shown in Figure 5.12a, the corresponding family in the frequency domain as a function of δ is given in Figure 5.21. These curves are again normalized against the naturalized frequency, $u = \omega/\omega_n$. Unlike the response curves (the normalized axis is $u = \omega/\omega_n$) time-domain step response, where the overshoot can reach twice the height of the final value of the step as $\delta \to 0$, the resonance peak in the frequency domain can approach infinity for no damping! The reader is assumed to have a good understanding of Bode plots along with the use of the second-order curves of Figure 5.21. However, a few crucial steps for Bode plotting will be reviewed. First, recall that the form of the transfer function is just the opposite as that for root locus. All of the pole and zero terms should be written in terms of time constants to give the correct values of K_{bode}. For the following transfer function as in Figure 5.22a, it should be rewritten for use in Bode plots as in Figure 5.22b.

Here, it should be noted that $K_{\text{Bode}} = Ka/(b\omega_n^2)$ and if $s \to j\omega \to 0$, then all terms enclosed in the parenthesis approach unity, and, of course, $G(j0) \to K_{\text{Bode}}$. The following plot (see Fig. 5.23) demonstrates the ease of making a Bode plot from the above equations.

The transfer function for the various time constants and the one second-order term as plotted (with $s \to j\omega$) is repeated here.

$$G(j\omega) = K_{\text{bode}}(\tau_a j\omega + 1)/\{(\tau_b j\omega + 1)[1 - (\omega/-\omega_n)^2 + 2j\delta\omega/-\omega_n]\}$$

$$\frac{K(s + a)}{(s + b)(s^2 + 2\zeta\omega_n s + \omega_n^2)}$$

$$\left(\frac{Ka}{b\omega_n^2}\right)\frac{[(1/a)s + 1]}{[s/b) + 1][(s/\omega_n)^2 + 2\zeta s/\omega_n + 1]}$$

(a) Root locus form

(b) Bode plot form

Figure 5.22 Different forms of G(s) for correct values of K

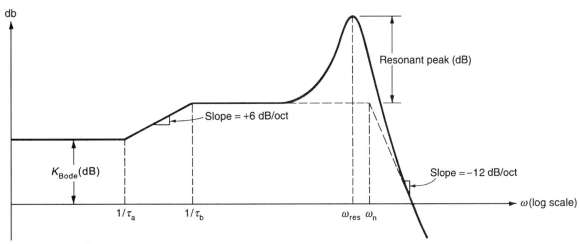

Figure 5.23   A bode plot (with a low ζ) of G $(j\omega)$

The slope of all asymptotes is an integer multiple of 6 dB/oct or 20 dB/dec (except near resonance for the second order term).

For another example, assume a frequency response test, obtained from laboratory data, is converted to dB and a smoothed plot is made on semi-log paper. From this plot (see Fig. 5.24), it is desired to determine the transfer function for the laboratory test device. Here, one would work backward to obtain the corner frequencies by laying out the asymptotes on the smoothed curve. At these corner frequencies the actual plot should deviate from the asymptotes by a "plotting error" of 3 dB for a first order term (recall, at $\omega = 1/\tau$, the parenthesis becomes $|(j1 + 1)|$ or $\sqrt{2}$ or 3 dB). The K_{Bode} (for convenience, the Bode subscript will be dropped) turns out to be equal to the intersection of the 0 dB axis and the extrapolated lowest frequency asymptote if the slope is +/− 6 dB/oct. (Recall, whatever the transfer function turns out to be, for $\omega \to 0$, any parenthesis terms become unity so that $|K/j\omega| = 1$ at 0 dB, then $K = \omega$.) It should now be clear that the numerical value of the transfer function, in terms of s, for Figure 6.24 is

$$G(s) = K(\tau_a s + 1)/[(sT_b + 1)(sT_c + 1)^2] = 2.5(s + 1)/[s(0.2s + 1)(0.05s + 1)^2].$$

If $j\omega$ is not first order, but is x-order, then the lowest frequency asymptote has a slope of $6x$ dB/oct and the intersection of the 0 dB axis would be $|K/\omega^x| = 1$ and

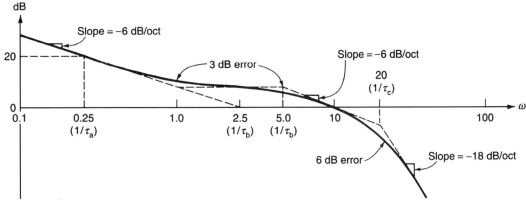

Figure 5.24   A Bode plot obtained from laboratory data

$K = \omega^x$. At the corner frequencies, the "plotting error" would be $3x$ dB for the x-order term.

From the test data, one could directly plot the phase vs frequency curve. However, the phase curve may be plotted indirectly. Recall, for each 6 dB/oct of slope of the magnitude curve, the corresponding phase curve has 90° phase. Thus the phase curve (of Fig. 5.24) would start with a −90°, change toward 0° at $\omega = 1$, change toward −90° at $\omega = 5$, and toward (and finish) −270° at $\omega = 20$ rad/second. An easy way to plot phase asymptotes is to assume a 45° phase shift at a corner frequency and 0° phase at $0.1\omega_{c.f.}$ and 90° at $10\omega_{c.f.}$; this approximation will not be in error more than 5.7° for first-order functions. Bode plots for stability, phase and gain margin, and so on, will be reviewed in more detail after the following short introduction to polar plots and Nyquist theory is presented.

Steady State Frequency Analysis—Polar and Nyquist

Plotting $G(j\omega)$ on a polar plane will lead directly to Nyquist theory. Recall, as the frequency is varied from zero to infinity, the polar plot is found from plotting the length of the phasor, the absolute value of $G(j\omega)$, at the phase angle for the particular frequencies being considered. The locus of these tips of the phasors produce this plot. Generally only a few frequencies near critical points need to be used. As another example, assume one wants to plot the transfer function of $G(j\omega) = K/[j\omega(\tau j\omega + 1)]$. Here, for very low frequencies, the singular $j\omega$ term in the denominator is very small and the angle is 90°, while the parenthesis term is near unity; therefore, the $G(j\omega)$ phasor is long and is pointing toward −90(+)°, while, for a very high frequency, the phasor length is very short and points toward −180(−)°. On the other hand if $\omega = 1/\tau$, then $G(j\Omega)$ becomes

$$G(j\omega)|_{w \to 1/\tau} = K/[j(1/\tau)(j1+1)] = (K\tau/\sqrt{2})\angle{-135°}.$$

A very rough curve may be plotted for three points; if $E_{in}(j\omega)$ has a value of one (either rms or peak—since the amplitude is a ratio) then $E_{out}(j\omega) = G(j\omega)E_{in}(j\omega) = G(j\omega)$. Consider Figure 5.25.

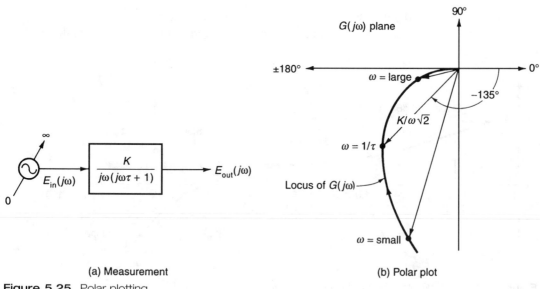

(a) Measurement   (b) Polar plot

Figure 5.25 Polar plotting

The polar plot data frequently is obtained by first plotting both a Bode magnitude (must compute actual magnitude from dB) and phase plot. However the data is obtained, a few guidelines for finding the polar path are helpful (for convenience, s, rather than $j\Omega$ will be used). If the order of the denominator is greater than that of the numerator of G, then the polar plot will always start from infinity (if a type 1 function, it will start from $-90°$, if a type 2, it will start from $-180°$, etc.) and will end up at the origin coming in at an angle equal to the numbers in the numerator minus the numbers in the denominator times $90°$. For a transfer function of

$$G(s) = K(\tau_a s + 1)/[s^2(\tau_b s + 1)^2]$$

the mapping from one plane to another is given in Figure 5.26a, b.

Here, the plot will include a reference length of -1; it will soon be shown that this point is the critical stability point.

The polar plot is as a result of changing the frequency from a very low value through a high one as in Figure 5.26b. This is the same as evaluating pole and zero locations in the s-plane for corresponding frequency points along the $+j\omega$ axis. If there were such a thing as a negative sine wave function generator, when changing this frequency through minus infinity up to zero the resulting plot would be the mirror image (or dashed line in Figure 5.26b). This would be the same as going along the $-j\omega$ axis in the s-plane.

From Nyquist theory, it is easy to show that if one were to enclose the entire RHP in the s-plane (the enclosure includes completing the path at a radius of infinity and bypassing any poles at the origin or on the $j\Omega$ axis at a small radius), the number of enclosures of the -1.0 point in the $G(j\omega)$ would indicate the number of poles and zeros in the RH s-plane. Thus, by plotting the open loop zeros and poles in the s-plane, one can determine if there are any closed loop poles in the RHP from examining the polar plot in the $G(j\omega)$ plane. Another example problem will be given after the following short discussion of the $F(j\omega)$-plane.

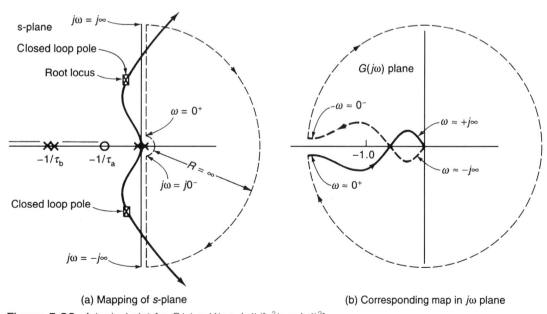

(a) Mapping of s-plane   (b) Corresponding map in $j\omega$ plane

Figure 5.26 A typical plot for $G(s) = K(\tau_a s + 1)/[s^2(\tau_b s + 1)^2]$

In a closed loop system, the denominator function, $F(s) = 1 + G(s)$, has a zero value if $|G(s)| = 1$ and if the angle of $\angle G(s) = +/- 180°$; the relationship in the $G(j\omega)$ plane is the -1.0 point, and, furthermore, the $F(j\omega)$ plane is located in the $G(j\omega)$ plane by merely shifting the axis over to the -1.0 point. Nyquist theory in essence states that the number of zeros minus the number of poles within a closed path in the $G(s)$ plane is represented in the $F(j\omega)$ plane by the number of encirclements of the origin. This statement may be interpreted by noting that any closed loop poles in the RHP (which represents an unstable system—unless countered by RHP zeros) will cause at least one encirclement of the -1.0 point in the $G(j\omega)$ plane. The encirclement is, of course, the mapping of the previously discussed path, starting from $j\omega = j0+$, all the way around the RH path in the s-plane.

For a short example, recall the root locus plot of Figure 5.18a (without the compensator) and also of Figure 5.27a for $G(s) = K/[s(s + 1)(s + 2)]$. For values of K of less than 6 (here, $K = K_{RL}$), all closed loop poles are in the LHP and the system is stable. However, if K is greater than 6, two of the closed loop poles will be in the RHP, resulting in an unstable system. Assume an open loop transfer function is plotted in the $G(j\omega)$ plane for two values of K (see Fig. 5.25b), the corresponding root locus plot (see Fig. 5.27a) has closed loop poles locations given for the two different values of K. Notice that in Figure 5.27b, the mapping indicates that the -1.0 point is enclosed twice for the larger value of K, while not at all for the smaller value. It is possible to choose a value of K so that the system is marginally stable (here, $K = 6$). From this concept, one may define gain margin, GM. Essentially, gain margin implies the closeness of the path to the -1.0 point; a simplified explanation (for a unity feedback path system) is that the **gain margin** may be defined as the numerical factor that one would have to multiply the original gain by to make the $G(j\omega)$ path just pass through the -1.0 point in the $G(j\omega)$ plane.

If one were to measure the distance, d, from the origin to where the path crosses the $180°$ axis (see Fig. 5.28a), then the inverse of this distance is the gain margin. Assume the distance d (for the smaller K) is 0.5 units, then GM $= 1/0.5 = 2.0$ (or could be converted to dB, here, 6 dB). If, on the other hand, one could put an ideal phase shifter, $e^{-j\Phi}$, in series with the transfer function, there would

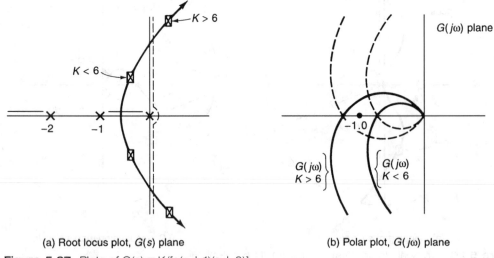

(a) Root locus plot, $G(s)$ plane   (b) Polar plot, $G(j\omega)$ plane

Figure 5.27   Plots of G(s) = K/[s(s + 1)(s + 2)]

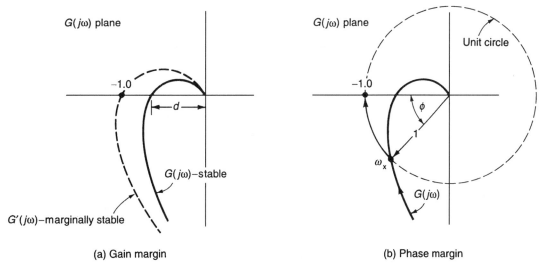

(a) Gain margin   (b) Phase margin

Figure 5.28 Gain and phase margin for a unity FB system

be a particular phase for which the length of the phasor, G($j\omega$), is unity; this quantity leads to the simplified definition of **phase margin**, PM, as "the amount of phase shift needed to modify the original plot, shifted by an amount Φ, such that the new plot passes through the –1.0 point."

In practice, the easy way to find phase margin is merely to scribe a unit circle around the origin of the $G(j\omega)$ plane. The intersection of this circle with the path of $G(j\omega)$ guarantees the tip of the phasor at that particular point to be unity length (the frequency of $G(j\omega)$ is also defined at this point). Also, the angle between the 180° axis and this particular phasor is Φ.

The rule of thumb is that the nearer to unity for gain margin and the smaller Φ or phase margin, the closer to instability the system becomes. There is a direct relationship between phase margin and the value of zeta, but this relationship is only for a second-order system, as is the step response. This relationship is nonlinear but is approximately given by $\Phi = 100\zeta$ (up to about 65°). For instance, for a ζ of 0.4, Φ is approximately 40° (the actual value for a second order system is 43°).

The concepts of phase and gain margin are easily displayed on an open loop Bode plot. From these plots, for a second-order function, one may predict the closed loop behavior of the system. However, even if the function is not second-order but does have a pair of dominant poles, the prediction is usually quite good. Actually the system behavior may be predicted for any order system by the use of a Nichol's chart, which may be found in almost any text on control system. To interpret the Bode plot (both magnitude and phase), it is convenient to plot a phase plot on the same plot as the magnitude (in dB) where the 0 dB axis is also the –180° phase axis. When the magnitude curve passes through the 0 dB axis, the magnitude of G is unity; this is the frequency point where the phase is measured to determine PM. On the other hand, where the phase curve passes through the –180° axis is the frequency where the GM is measured. A short example follows.

Assume the previous root locus problem of Figure 5.19a. For a zeta = 0.4, the gain was 1.35. The transfer function is repeated here and also is converted to Bode format as

$$G = 1.35/[s(s + 1)(s + 2)] = 0.675/[s(s + 1)(0.5s + 1)].$$

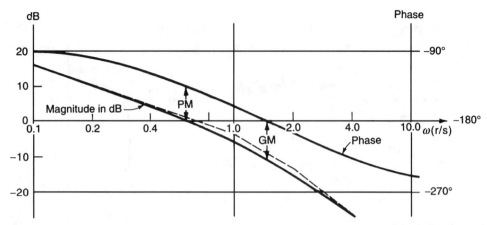

Figure 5.29 The gain and phase margin Bode plot

The Bode magnitude (in dB) and the phase (with the −180 axis corresponding to the 0 dB axis) is given in Figure 5.29.

From Figure 5.29, the GM is measured to be 13 dB at $\omega = 1.42$ r/s and PM is 44° at $\omega = 0.56$ r/s. Although these values were determined with a computer program, the Bode graphical procedure should be acceptable (within +/− several percent of these values) for examination purposes.

The reader may wish to use the same problem and add the phase lead compensator (*i.e.*, the modified transfer function) on the same Bode plot. It will be noted that the new gain and phase margins are approximately the same, but the cross-over frequencies are higher. The cross-over frequencies are a measure of the speed of response of the system. These frequencies will increase approximately as ω_n increases, as observed during the root locus compensation process.

A Modulated (2-Phase) System

These modulated systems are usually referred to as "suppressed carrier," or sometimes as "ac" systems. The analysis of these "ac" systems is mathematically much the same as the so-called "dc" ones in that the envelope (either peak or rms values) of the "ac", or carrier, is the signal that the Laplace transform represents. This is so if the signal frequency is much lower than the carrier. As an example, if a carrier is 60 Hz and is modulated at a signal frequency of 2 Hz, the envelope of the carrier is a sinusoidal signal of 2 Hz (actually, the frequency spectrum is given by two side-bands, one at 58 Hz, and the other at 62 Hz). Consider the simplest possible system of two potentiometers as error detectors connected to a two-phase motor through a power amplifier. Instead of an input of 2 Hz, assume a step input for θ_R of 1 radian (see Fig.5.30). Note that to determine if the envelope input to the motor is plus or minus, it needs to be referenced to the carrier (that is, is it in-phase, or is it 180° out-of-phase with respect to the carrier?). Here the motor's main windings act to detect (or to demodulate) the carrier. Also note that if one observed the error signal in an oscilloscope, then the peak overshoot, the time-to-the-first-peak, the "ringing" frequency may be obtained for the system if one could imagine that someone could sketch the envelope with a felt pen on the face of the screen as the signal was obtained. An open loop frequency response test from a very low frequency (say from 0.01 Hz) to a high frequency (say 5.0 Hz, but well below the carrier frequency). This information is readily converted to a Bode plot or a Nichol's chart to predict the closed loop system performance.

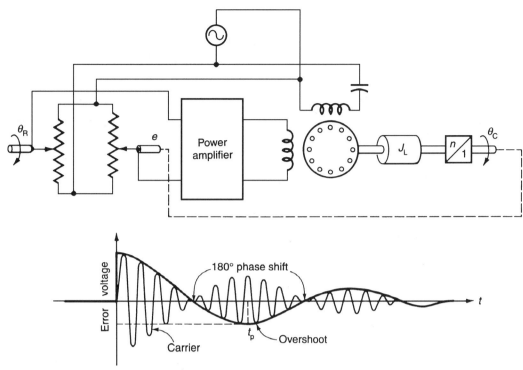

Figure 5.30 An "ac" system using a two-phase servo motor

Advanced Methods of Analysis (for reference only)

State Variable Conversion

To convert from a transfer function format to that of a state variable format, one must first recall the meaning of a state variable. Consider a standard second-order system excited by a step input.

$$\text{Step} \longrightarrow \boxed{\dfrac{K}{s^2 + as + b}} \longrightarrow \theta_0$$

The response, θ, as a function of time for this type of function is well known and is repeated here for clarity. If one were to replot this function as the derivative— say $\dot{\Theta}$, or ω, vs. the function itself (*i.e.,* ω vs. θ) where time is now implicit, the plot would be known as a "phase plane" (see Fig.5.31b).

It is usually more convenient to redefine the variables as x_1 and x_2 (*i.e.,* $x_1 = \theta$ and $x_2 = \dot{x}_1 = \omega$), then the plot may be considered as the locus of the coordinates x_1 and x_2. The locus of the vector coordinates as a function of time is then the state variable (see Fig. 5.32), $x(t)$.

Consider the above second-order system using "mixed notation" (*i.e.,* for the Laplace $sX_1(s)$ written as $\dot{x}_1 = x_2$ in the time domain—if one first considers that all initial conditions are zero) and then write the differential equation (from which the transfer function came from) as follows:

$$\ddot{\Theta} + a\,\dot{\Theta} + b\,\Theta = Ke_{in} \qquad \dot{x}_1 = x_2$$

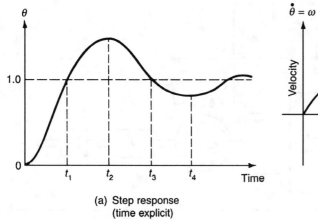

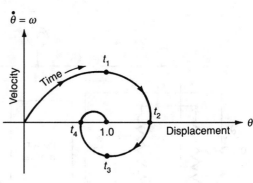

(a) Step response
(time explicit)

(b) Phase plane
(time implicit)

Figure 5.31 A phase plane plot for a step input

then

$$x_2 + ax_2 + bx_1 = Ke_{in} = Ku \qquad x_2 = -ax_1 - bx_2 + Ku$$

The mixed notation block diagram equivalent is shown in Figure 5.33 and the matrix form of the equation is given by (the underlined characters are usually written as bold faced in most texts)

$$\left. \begin{array}{l} \dot{x}_1 = x_2 \\ \dot{x}_2 = -bx_1 - ax_2 \end{array} \right\} \quad \text{or} \quad \left\{ \dot{\underline{x}} = \begin{bmatrix} 0 & 1 \\ -b & -a \end{bmatrix} \underline{x} + \begin{bmatrix} 0 \\ K \end{bmatrix} u \right.$$

It follows directly from the equations that the equivalent state variable block diagram is as shown in Figure 5.34. where

$$A = \begin{bmatrix} 0 & 1 \\ -a & -b \end{bmatrix}$$

$$B = \begin{bmatrix} 0 \\ K \end{bmatrix}, C = [1 \quad 0]$$

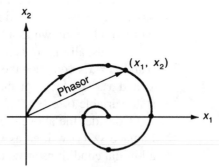

Figure 5.32 For any particular time, the vector coordinates give the state of the system

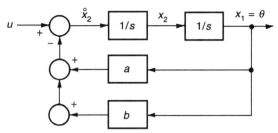

Figure 5.33 Mixed block diagram

Consider the following numerical example:

$$U(s) \longrightarrow \boxed{\dfrac{K}{s^2(s+1)^2}} \longrightarrow \theta(s)$$

Temporarily assume all initial conditions are zero; the differential equation that the transfer function originally came from may be written as

$$2\Theta + \ddot{\Theta} = Ku$$

Define $x_1 = \Theta$, $x_2 = \dot{x}_1 = \dot{\Theta}$, etc.

$$\dot{x}_4 + 2x_4 + x_3 = Ku$$

Then, again, the mixed notation block diagram may be written as in Figure 5.35 and the state variable matrix form of the equation is (please refer to Fig. 5.34),

$$\underline{A} = \begin{bmatrix} 0 & 1 & 0 & 0 \\ 0 & 0 & 1 & 0 \\ 0 & 0 & 0 & 1 \\ 0 & 0 & -1 & -2 \end{bmatrix} \quad \underline{B} = \begin{bmatrix} 0 \\ 0 \\ 0 \\ K \end{bmatrix} \quad \underline{C} = [1 \quad 0 \quad 0 \quad 0]$$

For a unity feedback path around the transfer function, the form of the state variable block diagram is the same except the matrix constants change as follows in Figure 5.35 (recall the rule about post- and pre-multiplying when manipulating matrices.)

$$\dot{\underline{x}} = \underline{A}\underline{x} + \underline{B}u = \underline{A}\underline{x} + \underline{B}(r - 0)$$
$$= \underline{A}\underline{x} + \underline{B}(r - \underline{C}\underline{x}) = (\underline{A} - \underline{B}\underline{C})\underline{x} + \underline{B}r$$
$$\dot{\underline{x}} = \underline{A}_{C1}\underline{x} + \underline{B}r.$$

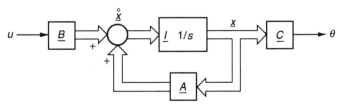

Figure 5.34 State block diagram

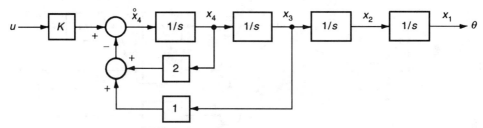

Figure 5.35 Mixed notation block diagram for the example

The methods of solution and the operations on the equations may be found in almost any book on control systems. However, the reader must recall the meaning of solution(s); if he finds the vector x at any time, say t_1 (*i.e.,* "freeze" the vector at t_1) then x's are merely the coordinates of the vector in the phase plane (which is easy to visualize for a phase plane of a second-order system). These coordinates are the initial conditions if one were to "unfreeze" time and start up again.

Although the probability of in-depth questions on state variables may be minimal, it is recommended that one should at least review how to convert to the state variable format and then how to solve for their solutions, by either time domain or Laplace techniques.

Many simulation computer programs are available for the conversion of transfer function format to the state variable presentation. Of course, these computer programs also allow one to solve directly for various manipulations and the solution. Because one does not have access to these simulation programs on an examination, it is suggested that one review pencil and paper techniques on simpler problems. (You may be expected to answer questions on the subject of simulation itself for the problems in the area of computers.)

Sampled Data Systems and the *z*-Transform (Optional Information)

For sampled systems (usually used with digital computer control) one must recall that any time a signal is held or delayed, the effect is much the same as inserting the delay function, e^{-sT}, into a typical Laplace transfer function model. Thus, to gain a better feel for the behavior of a computer-controlled system, consider a computer algorithm that does nothing to a signal except pass it through the computer and associated sample and hold circuit and A/D and D/A converters (for lack of a better name, use "unity thru-put"). Then the model may be thought of as in Figure 5.37.

An error signal wave form for a typical second-order system due to a step input might have the shape of Figure 5.38.

The approximated signal at $m(t)$ is the same as that at $e(t)$ except delayed by a phase shift of $T/2$ (or one half the time of the delayed steps). From classical

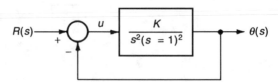

Figure 5.36 A unity feedback system

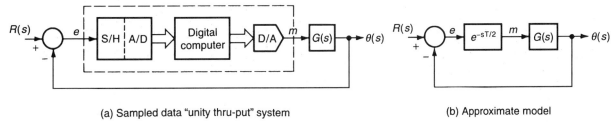

(a) Sampled data "unity thru-put" system　　　　(b) Approximate model

Figure 5.37 A sampled data system

theory, one may recall the effect of time delay with regard to stability by use of the Nyquist plot (see Fig. 5.39).

Obviously, the system tends toward instability for greater delay and may need compensation. The Laplace delay function, e^{-sT}, is very similar to the effect of z^{-1} when using z-transforms. Thus a short review of z-transforms is in order.

Recall that a sampling switch is assumed to close only for an instant of time; then, whatever signal is sampled, the signal is held until the next sample time. If the Laplace of the unsampled signal is $E(s)$, then the delayed sampled signal may be given by

$$\mathcal{L}[f^*(t)] = \sum_{k=0}^{\infty} f(kT)e^{-kTs} = \sum_{k=0}^{\infty} f(\mathrm{k T})z^{-k}, \quad \text{where } z = e^{sT} \qquad \textbf{(5.12)}$$

Then direct substitution of a particular function, say a step, into the equation would yield the z-transform for that function.

$$z[u(t)] = \sum_{k=0}^{\infty} \mathrm{u}(kT)z^{-K} = \sum_{k=0}^{\infty} z^{-k} = \frac{z}{z-1}$$

where z^{-k} may be represented by its geometric infinite series as a closed function of

$$z^{-k} = 1 + z^{-1} + z^{-2} + z^{-3} + \cdots = z/(z-1). \qquad \textbf{(5.13)}$$

The z-transform for several functions are shown in Table 5.2.

The sample and hold (with the time delay) has, as its Laplace representation, the effect of

$$\mathcal{L} \ [\text{o——o} \diagup \text{o}^{\;T} \ \boxed{\text{S/H}} \ \text{——o}] = (1 - e^{-sT})/s$$

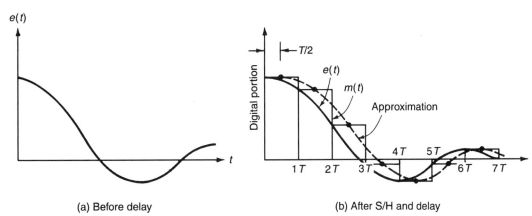

(a) Before delay　　　　　　　(b) After S/H and delay

Figure 5.38 An error signal converted from analog to S/H

Table 5.2 z-transform for several functions

	$f(t)$	$F(s)$	$F(z)$
Step	$u(t)$	$1/s$	$z/(z-1)$
Ramp	$tu(t)$	$1/s^2$	$Tz/(z-1)^2$
Exp	e^{-at}	$1/(s+a)$	$z/(z-e^{-aT})$
Sine	$(\sin \omega t)u(t)$	$\omega/(s^2+s^2)$	$z \sin \omega T/(z^2 - 2z \cos\omega T + 1)$

And the method of analysis, if the transfer function (in "s") is known, is to make the whole system as a Laplace transform, *then convert to the z-transform as a whole.* The following simple example will illustrate the method:

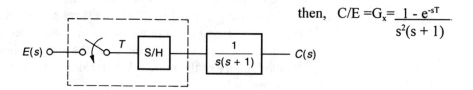

then, $\quad C/E = G_x = \dfrac{1 - e^{-sT}}{s^2(s+1)}$

To then convert to the z-transform, merely expand the total Laplace transform over the partial fraction expansion and convert directly to the desired result (recall, $e^{-sT} \equiv z^{-1}$).

$$G_x(s) = (1 - e^{-st})[1/s^2 - 1/s + 1/(s+1)] \rightarrow (1 - z^{-1})[Tz/(z-1)$$
$$- z/(z-1) + z/(z-e^{-T})]$$

One may find the response to a unit impulse input exactly the same as the method for the Laplace transform technique (*i.e.,* multiplying the functions in "z"-together).

$$C(z) = [G(z)](1) \quad \text{since} \quad [\text{Impulse}] = 1$$

Or, for a numerical value, assume the sample period is known to be one second. Then, simplifying,

$$C(z) = [(ze^{-T} - z + Tz) + (1 - e^{-T} - Te^{-T})]/[(z-1)(z-e^{-T})]|_{T=1}$$
$$= (0.368z + 0.264)/(z^2 - 1.368z + 0.368).$$

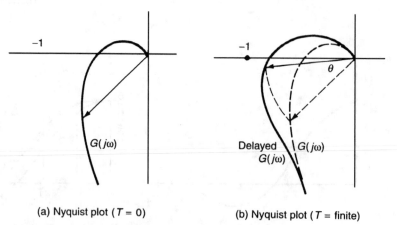

(a) Nyquist plot ($T = 0$)   (b) Nyquist plot ($T =$ finite)

Figure 5.39 A Nyquist plot before and after time delay

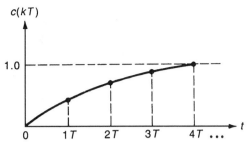

Figure 5.40 The sampled response for the example problem

Finally, the answer for the step response may be found by several different methods; perhaps the easiest way is merely to divide the numerator by the denominator.

$$z^2 - 1.368z + 0.368 \overline{)\ 0.368 + 0.264} \quad \begin{array}{l} 0.368z^{-1} + 0767z^{-2} + 0.914z^{-3} + 0.968z^{-4} + \cdots \end{array}$$

The response (in "z") is directly converted to a difference equation (in "k") and gives the result at discrete times (with no information about points between the sample periods, kT).

$$c(kT) = 0.368(k = 1) + 0.762(k = 2) + 0.914(k = 3) + 0.968(k = 4) + \cdots$$

This effect is shown in Figure 5.40.

For a closed loop system, the method is the same (see Fig. 5.41).

To find whether the system is stable, recall that for a system in "s," if there are no poles in the right half plane, the system is stable. That is, the "$j\omega$" axis is the dividing line. Consider the equation e^{-sT}, where $sT = aT + j\omega T$; $j\omega T$ represents the angle, and e^{aT} represents the magnitude of z. The dividing line is thus when alpha equals zero, or the absolute value of z is unity. Thus, the $j\omega$ axis is the s-plane, equivalent to the unit circle in the z-plane, as shown in Figure 5.42.

The stability of the system may easily be determined by calculating the poles of the characteristic polynomial (CP) of the function (in z), and determining if they are *within* the unit circle (again assume T = one second).

$$1 + G(z) = 1 + (0.368z + 0.264)/(z^2 - 1.368z + 0.368)$$
$$= (z^2 - z + 0.632)/(z^2 - 1.368z + 0.368)$$
$$\text{CP} = z^2 - z + 0.632 = (z - 0.5 + j0.618)(z - 0.5 + j0.618)$$

For the closed loop system (assuming the sample period is unity), the characteristic polynomial has roots within the circle and is therefore stable. One should note that stability (for a linear system) does not depend on the input.

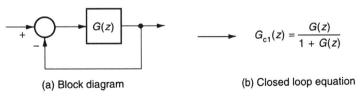

(a) Block diagram   (b) Closed loop equation

Figure 5.41 The closed loop relationship for H = unity

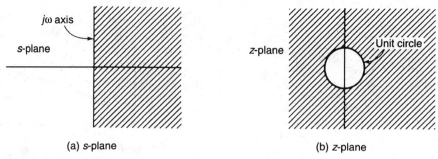

(a) *s*-plane   (b) *z*-plane

Figure 5.42   Stability limits for both the s- and z-plane

Example Question

Repeat the preceding sample problem (in closed loop) but let the input be a unit step. Also, find whether the system is stable.

Solution

The same method as before, except the system is excited by the step, $z/(z-1)$, rather than unity. The output equation (in z) becomes (again for T = unity)

$$c(kT) = 0.368(k = 1) + 1.0(k = 2) + 1.4(k = 3) + 1.4(k = 4)$$
$$+ 1.15(k = 5) + \cdots$$

For stability, the answer is the same as before because stability is independent of the excitation.

This short review presented on sampled data systems is somewhat unconventional. Certain approximations have been used to condense the subject matter to a few pages and hopefully to give the reader a "feel" for these kinds of systems.[*]

RECOMMENDED REFERENCES

Franklin, Powell, and Emami-Naeini, *Feeback Control of Dynamic Systems,* 4th Edition, Prentice Hall, 2002.

Jones, Lincoln, "Digital Control of Analog Systems," printed in the Annual Conference Proceedings of the American Society of Engineering Education (ASEE, 1986).

---

[*] For more detailed information on this z-transform approximation technique, the reader may refer to a paper by this author, *Digital Control of Analog Systems*, printed in the Annual Conference Proceedings of the American Society of Engineering Education (ASEE, 1986).

Computing

OUTLINE

This chapter deals with computing in the areas of analog, digital, and "packaged" digital programs. The material under the heading "computer systems" is extremely broad; here it will be limited to uses in engineering design and analysis (rather than the software design for the programs themselves). While analog computers are now rarely used in practice, some analog concepts involving integro-differential equations for formulating computer solutions to engineering problems are important in understanding digital packaged programs. Therefore this chapter includes a brief review of analog computing fundamentals.

In this author's opinion, the writers of the professional examination questions are more interested in determining the examinee's fundamental knowledge of computing principles rather than his or her familiarity with the latest programming language. Consequently only major concepts are presented in this chapter. Short examples covering selected kinds of computing are also included.

Most universities require their electrical engineering graduates to be familiar with one or more digital computer circuit design packages (such as PSpice for general circuit design and analysis). Others are more aggressive and require students to be proficient with a number of specialized programs. These also cover specific applications such as control systems (ACSL, CSMP, MATRIXx, CSSL, CC, TUTSIM, etc.). Some form of digital computing may be expected in almost all specialized areas.The brief review of certain concepts from analog computing is followed by a review of writing or using digital programs.

ANALOG COMPUTING FUNDAMENTALS

These brief notes on the theory of analog computing do not include details or instructions for actual operation nor do they go into the details of time or magnitude scaling (as needed for actual computing). However, enough material is given to aid the reader in understanding the basic concepts and problem-solving methods of analog computing.

For this discussion, analog computing concepts are used for problems that can be described by differential equations (DE) or by Laplace transfer functions. One concept will be as shown in Figure 6.1.

An analog computer consists of a collection of operational amplifiers and related switching circuits and components. The heart of the computer may be considered to be a group of operational amplifiers connected in certain configurations to give elements that act like integrators, summers, inverters, and other specialized units. These elements are interconnected to allow for the solution of both linear and nonlinear differential equations.

Operational amplifier configurations for use in computing circuits are assumed to be in the inverting mode (*i.e.,* positive differential input to ground and signal input on the negative input). Also, each op-amp is considered ideal in that its input impedance is infinite ($>10^8 \, \Omega$) and its output impedance is zero ($<50 \, \Omega$). An important practical limitation is that all op-amps are operated in their linear range. Because of the brevity of this review, the techniques for guaranteeing this requirement are ignored.

Use of Operational Amplifiers in Computing Circuits

All computing circuit elements will have interconnecting switching circuits so that three modes of operation may be obtained. These modes are

1. INITIALIZATION or RESET (for setting any initial conditions),

2. COMPUTE (actively computing), and

3. HOLD ("freezing the solution" at any particular instant of time).

For the interested reader, the circuitry for these modes will be addressed later.

A circuit configuration that allows for inverting and multiplying by a constant is shown in Figure 6.2. (It was customary during the analog computer era to use a rounded head for symbol for the open loop op-amp.) Figure 6.2 shows that the output voltage is given as $e_{\text{out}} = -(R_f/R_1)e_1 = K_1 e_1$. The key to this relationship is that the voltage at the input to the op-amp, e_j, is very near zero (a few microvolts) as compared to perhaps several volts at e_1 or e_{out}. A constant multiplier is given as K_1. A modifiable multiplier could be introduced by adding a potentiometer into the circuit. This would serve as a voltage divider with negligible resistance (see Fig. 6.3).

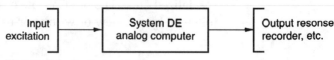

Figure 6.1   System model concept

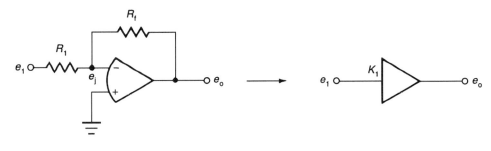

(a) Basic configuration   (b) Representation

Figure 6.2 Inverting circuit and its representation

If this circuit is to be used as a SUMMER, the output voltage is given as

$$e_{\text{out}} = -(R_f/R_1)\, e_1 - (R_f/R_2)\, e_2... - (R_f/R_n)e_n. \tag{6.1}$$

A simple example problem for a summing circuit is shown below. The example is based on the following relationship:

$$W = 2X + 3.7Y - 0.23Z$$

Let $W = -e(\text{out})$, $X = e_1$, $Y = e_2$, $Z = e_3$, and $K_1 = 2$, $\beta_2 K_2 = 0.37x10$, $\beta_3 K_3 = 0.23x1$. The resulting circuit schematic is shown is Figure 6.4.

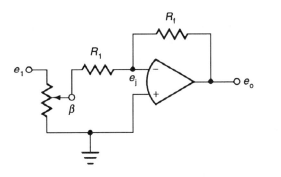

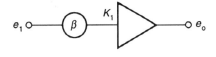

(a) Wiring schematic   (b) Computer schematic

Figure 6.3 Multiplying by a modifiable multiplier

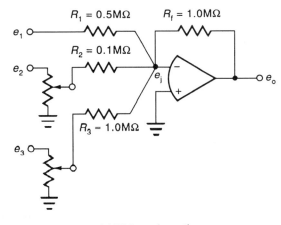

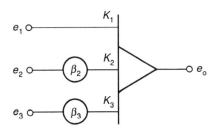

(a) Wiring schematic   (b) Computer schematic

Figure 6.4 Circuit diagram for example problem

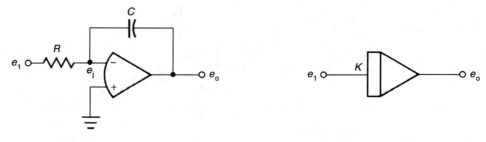

(a) Circuit wiring schematic   (b) Computer schematic

Figure 6.5   Integration where $K = 1/RC$. The R's may have units of megohms and C's may have units of microfarads. They will therefore cancel

The operational amplifier with a capacitor in feedback is an ideal device to use for INTEGRATION (recall, $i = dq/dt$ and $v = q/C$). The relationship for integration is as follows (for zero Intitial Conditions, IC's):

$$v_C = 1/C \int_{-\infty}^{+\infty} i\, dt \qquad (6.2)$$

The actual circuit configuration is given in Figure 6.5. It is assumed that both e_j and i_j are very small and may be neglected.

One should realize that these active circuits give almost perfect integration and should not be confused with passive approximation techniques. For integration where there is an initial voltage across the feedback capacitor—BEFORE INTEGRATION STARTS—these circuits exhibit an initial voltage at the output (as shown in Eq. 6.3),

$$e_o = -K \int_0^\infty e_1\, dt - K \int_\infty^0 e_x\, dt = -K \int_0^\infty e_1\, dt - V_{ic} \qquad (6.3)$$

It is also possible to sum on an integrator. The equations for summing are self-evident but are included for completeness.

$$e_o = -(1/C) \int_0^\infty [(1/R_1)e_1 + (1/R_2)e_2 + \cdots (1/R_n)e_n]\, dt + V_{ic} \qquad (6.4)$$

Consider the following network shown in Figure 6.6.

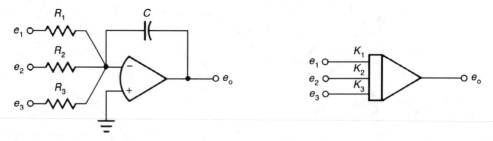

(a) Wiring schematic   (b) Computer schematic

Figure 6.6   Summing on an integrator

Differentiation circuits are not shown because they present many problems for the circuit designer (the same is true for digital computing). Theoretically, interchanging the R and the C in the integration circuit will yield a differentiation circuit. In practice, however, randomly generated noise will saturate the op-amp in the circuit so they are usually not used. There are a number of approximation circuits that may be used in lieu of this configuration.

Solving Differential Equations

Consider the following first-order differential equation with zero initial conditions:

$$a\left(\frac{dx}{dt}\right) + bx = c \implies a\dot{x} + bx = c, \, x(0) = 0.$$

The solution is straightforward. Solving by hand for the highest derivative gives

$$\dot{x} = -(b/a)x + (1/a)c$$

$$e_o = -(\beta_2 K_2)e_2 + \beta_1 K_1 e_1$$

If this solution is incorporated into a summer, the output will be the negative of the first derivative. Because x is the full integral of x-dot, applying the summer's output to the input of the integrator will yield the required solution (see Fig. 6.7). The paradox of needing the negative of the solution to actually "compute" the solution has a simple explanation. The only requirement for solving a first-order differential equation is that the initial condition be known. In the current case, it is assumed to be zero. If the initial charge on the capacitor is zero before the computer button is pressed, the known summer output (zero, in this case) is inverted, and is fed back to the integrator's input as in Figure 6.8.

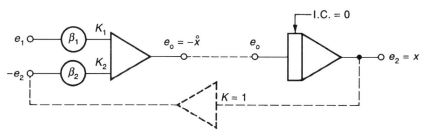

Figure 6.7 First-order differential equation solution

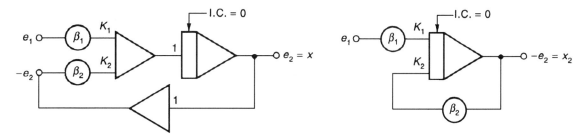

(a) Complete schematic

(b) Simplified schematic

Figure 6.8 Alternate first-order differential equation solution

Initial Conditions

The practical aspect of obtaining initial conditions involves the internal and external connections of the circuit and the method a particular manufacturer might choose to arrange the switches for the OPERATION or COMPUTE mode. Usually, the internal switches are simply a way of applying a charging voltage to the integrating capacitors. Before t = 0 (*i.e.,* while the integrator is in the RESET mode), the circuit can be considered to have some desired voltage applied to the capacitor. The actual circuit for obtaining this voltage is usually similar to that shown in Figure 6.9. Because the op-amp input is essentially at ground potential, the output is then the initial condition. Also, the output voltage is the negative of the input voltage measured at the arm of the reference voltage potentiometer. The switch positions normally correspond to three commands: RESET, OPERATE (or COMPUTE), and HOLD (or "do nothing"). By use of initial conditions, the analog computer may be set to solve complex system models. These models are described by sets of differential equations exhibiting many different conditions.

Transfer Functions

Transfer functions, using Laplace transforms, can be converted to a form that is compatible with an analog computer. To perform the transformation, the original differential equation that the transform represents is recovered. Consider the following equivalents (assuming initial condition of zero and dropping the (*s*) notation):

$$E_o(s)/E_{in}(s) = K/(s\tau + 1) \implies s\tau E_o + E_o = KE_{in}$$

$$e_o = -(1/\tau)e_o + (K/\tau)e_{in}$$

The equivalent block diagram relationship (Fig. 6.10) has a corresponding analog configuration as shown in Figure 6.8.

Consider a system that includes feedback (derivative) compensation. This configuration is illustrated in Figure 6.11.

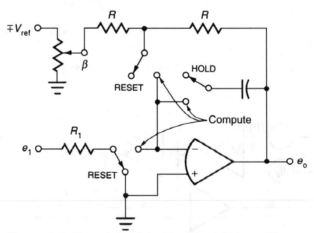

Figure 6.9 Actual circuit for obtaining initial conditions

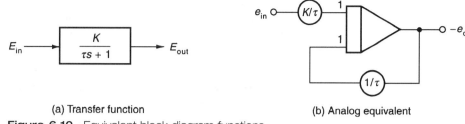

(a) Transfer function   (b) Analog equivalent

Figure 6.10 Equivalent block diagram functions

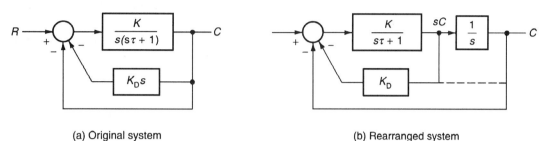

(a) Original system   (b) Rearranged system

Figure 6.11 A control system with derivative compensation

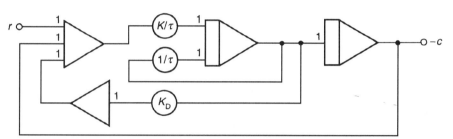

Figure 6.12 Analog computer schematic for Figure 6.11a

The analog computer configuration for the closed loop system has been slightly rearranged such that the derivative function is not needed. The signal before the last integrator is the derivative of the output (Fig. 6.12). This technique is used for digital computer programs as well.

Very complex system transfer function problems can be simulated on either the analog computer (especially if no time or magnitude scaling is involved) or, by using simulation packages, on the digital computer.

DIGITAL COMPUTING

The amount of material concerning digital computing has grown so rapidly that only a limited review is practical. The material in this section will therefore be general and is intended to serve only as a memory refresher.

Of the various languages, probably BASIC is the easiest language to review in a short amount of time. It is suggested that the reader obtain a quick reference booklet in order to review the purpose and syntax of the BASIC commands, statements, functions, and variables. Many of the concepts from one computer language implementation are applicable in other languages. If the reader is more familiar with JAVA, FORTRAN, C, PASCAL, etc., then one of these would be more appropriate for review. For simulation application, C is the preferred base language and is appropriate for review. Programs can be written for almost any problem. However, it is usually better to use a packaged program if one is available for the required application. Programs such as MATLAB® or Mathcad are also appropriate for technical computation.

For example, consider the halfwave rectifier circuit that converts ac voltage to dc voltage. It consists of an ac voltage source, a diode in series with a capacitor, which is in parallel with the load resistor. The voltage of the source is $V_s = V_o \sin(wt)$, where w is the radial frequency. The voltage across the resistor is $V_r = V_o \sin[wt]$ and $I_r = V_o \sin[wt]/R$. The current in the capacitor is $V_r = V_o wCV_o \cos[wt]$.

The following MATLAB program plots the voltage across the resistor V_r and the source V_s as a function of time, for times between 0 seconds and 70 msecs. To examine the effect of the capacitor size on the voltage across the load, utilize the capacitor values of 45 microfarads and 10 microfarads.

```
V0 = 12; C = 45E-6; R = 1800; F = 60;

RF = 70E-E;   W = 2*PI*F;

CLEAR T VRT VS

T = 0:0.-5E-3:TF;

N = LENGTH (T);

STATE = 'ON';

FOR I = 1:N

  VS(I) = V0*SIN(W*T(I));

  SWITCH STATE

    CASE 'ON'

    VR(I) = VS(I);

    IR = VS(I)/R;

    IC = W*C*V0*COS(W*T(I));

    SUMI = IR + IC;

IF SUMI <= 0

    STATE = 'OFF';

    TA = T(I);

  END

CASE = 'OFF'
```

```
VR(I) = V0SIN(W*TA)*EXP(-(T(I) - TA)/)R*C);

  IF VS(I) >= VR(I)

    STATE = 'ON';

    END

  END

END

PLOT(T,VS,':',T,VR,'K','LINEWIDTH',1)

XLABEL('TIME(S)');YLABEL('VOLTAGE(V)')
```

Integration Routines

Digital simulation programs normally require integration routines as well as other arithmetic operations. Numerical integration methods are used to create integration routines. One of the simplest of these routines is based on Euler's method. Euler's technique is not necessarily the best because it requires small step size to produce results that are reasonably accurate. Euler's methods are useful for providing insight into integration routines, and are briefly discussed below.

Integration routines produce values of system variables at specific points in time and update this information at each interval of delta T (here delta $T = t_{n+1} - t_n$). Instead of a continuous function of time, $x(t)$, the variable x will be represented with discrete values such that $x(t)$ is represented by $x_0, x_1, x_2, ..., x_n$. Consider a simple differential equation, as

$$dx/dt + ax = f(t), \quad \text{or}, \quad x + ax = f(t). \qquad (6.5)$$

It can be shown for $f(t)$ = unit step, discrete values will result in Equation 6.5 being approximated by

$$dx_n/dt = (x_{n+1} - x_n)/(t_{n+1} - t_n) = (x_{n+1} - x_n)/T \qquad (6.6a)$$

$$x_{n+1} = x_n - Tax_n + T. \qquad (6.6b)$$

By knowing the first value of x_n (or the initial condition), the solution may be achieved for as many "next values" of x_{n+1} as desired for some value of T. The difference Equation 6.6a is the approximate equivalent of the differential Equation 6.5 for small values of T. Although the solution of Equation 6.5 can be obtained directly on an analog computer, the difference equation may be programmed in most high-level languages on a digital computer. Consider the following differential equation:

$$x + 4x = \text{unit step input}, \quad x(0) = 0.$$

The FORTRAN solution[1] of this equation is given in Figure 6.13, for several values of step size $T(T = 0.1, 0.2, \text{and } 0.3)$.

---

[1] This program is a modification of one developed in a text by Bennet, *Introduction to Computer Simulation*, West Publishing Company, 1974, p. 379.

```
C  PROGRAM FOR SOLVING XDOT + AX = AX = 1.0
   REAL TI(20), X(20)
C  SET INITIAL VALUE OF X(0)
   X0 = 0.0
C  SET VALUE OF A
   A = 4.0
C  FINISH TIME EQUALS 5 TIME CONSTANTS
   FT = 5*(1/A)
C  FIND RESPONSE FOR 3 DIFFERENT VALUES OF T
   DO 10 J = 1, 3
   FJ = J
   T = FJ*0.1
C  DETERMINE THE NUMBER OF STEPS
   N = FT/T
C  CALCULATE FIRST VALUE OF X AT N + 1
   X(1) = X0 - T*A*X0 + T
C  CALCULATING THE VALUE OF X(I) AT EACH INCREMENT
   DO 20 I = 1, N
   AI = I
   X(I + 1) = X(I) - T*A* X(I) + T
   TI(I) = T*AI
20 CONTINUE
C
   WRITE(6, 30)
30 FORMAT(1X, 4H TIME, 4X, 1HX)
   WRITE(6, 40) TO, X0
40 FORMAT(1X, F4.2, F7.3)
   WRITE(6, 50) (TI(I), X(I), I = 1, N)
50 FORMAT(1X, F4.2, F7.3)
10 CONTINUE
   STOP
   END
```

Figure 6.13   FORTRAN program for solving: $x + 4x = 1$

Figure 6.14 is a comparison plot of the exact and approximate solutions. The exact solution is represented by the solid line while the computer solution for $T = 0.2$ second is given by "*" points. The solution produced using a stop interval of $T = 0.3$ second is shown with o's. The comparative plot shows that the smaller value of T yields a good approximation while the larger value gives an unacceptable solution. The time constant for the differential equation is 0.25 seconds. As a general rule, a T value should be used that is less than the shortest time constant of the equation being solved ($T = 0.1t$ is a recommended value). A step interval that is too small may result in an unacceptable number of iterations before the solution is reached.

Packaged Programs

Most currently available packaged simulation programs use algorithms based not on Euler's methods but on more advanced methods such as Range-Kutta's. Automatic variable step size methods like Milne's are frequently used. However, as mentioned before, these routines are all built into the packaged programs and may be transparent

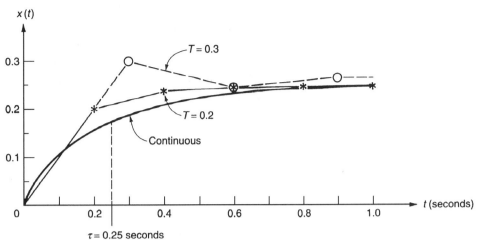

Figure 6.14 Plots of the solution of $x + 4x = 1$

to the user. The user of a specialized program may be without knowledge of the high-level language being employed (except for certain modifications). More discussion of these algorithms follows in the section on CSMP programming. For circuit analysis programs such as SPICE, these kinds of algorithms are also transparent to the user.

CIRCUIT SIMULATION—SPICE

Of the number of circuits programs available, the best (in this author's opinion) for solving a large number of electrical problems is SPICE (or PSpice). The acronym, SPICE, stands for Simulation Programs with Integrated Circuit Emphasis. Many commercially available programs for electrical circuit simulation have much in common with SPICE (although various calls and structure may be different). A smaller version of SPICE has been written for the IBM PC. The program is capable of solving problems involving dc and ac circuits, both independent and time-dependent sources, transient and steady-state solutions, Bode plots, sub-circuits, and other problem types. Only a few operations are reviewed here.

As with most programs, SPICE requires a source file that must be created before the program can analyze a particular electrical circuit. This file must do, at a minimum, three things: (1) it must describe the circuit (generally, in the form of node equations), (2) it must describe the kind of analysis that is to be performed (dc, steady-state ac, transient, frequency response, etc.), and (3) it must define the form of the output (*i.e.*, a graphical solution, a printed, output, etc.). In all cases, the source file must start (first line) with a title statement, and must end with an .END statement (the last line must start with a period). The descriptive item (1) is referred to as the data statement. It must follow a prescribed format and use numbers for symbols (*i.e.*, 5 kilo-units = 5E3).

The kind of analysis used (item 2), requires a group of control statements (control statements begin with a period except for a dc circuit analysis—dc is assumed to be a default case and is always included in an analysis); these statements allow one to state how to control various computations (that is, to call for a transient analysis with step size, start and stop values, etc.). The form of the output (item 3) desired may be specified with a .PRINT or a .PLOT statement (preceded with a period), including the conditions for these operations.

As previously mentioned, the circuit configuration for SPICE must be presented in terms of nodal analysis; each node must be assigned a number with the reference node being zero. Each node is required to be connected to another node through a described circuit element. Each of these elements must start with a prescribed letter (x's = any user defined alphanumeric identifier) as indicated in the following table:

Element	Spice Name
Resistor	Rxxxxxxx
Inductor	Lxxxxxxx
Capacitor	Cxxxxxxx
Independent Voltage Source	Vxxxxxxx
Independent Current Source	Ixxxxxxx
Voltage-controlled Voltage Source	Exxxxxxx
Voltage-controlled Current Source	Gxxxxxxx
Current-controlled Voltage Source	Hxxxxxxx
Current-controlled Current Source	Fxxxxxxx

To describe an element in a file, start with the name, give the node-to-node location, and specify the numeric value of the element (for an energy storage element, any initial conditions must also be included). If the element is to be an independent source, then whether it is dc or ac must also be included (after the nodes are specified). If the element is a dependent source, then the controlling nodes must be included (again, after the nodes are specified). For any independent sources or initial conditions, polarity should be noted. For a voltage source, the first named node is positive with respect to the second node; for the current source, the direction of current is from the first to the second node specified. A peculiarity of SPICE is that an independent voltage, assigned a value of zero, will yield a numeric value of current through this zero source.

A simple dc circuit will be given that contains both an independent voltage source and a dependent current one. Consider the circuit of Figure 6.15,

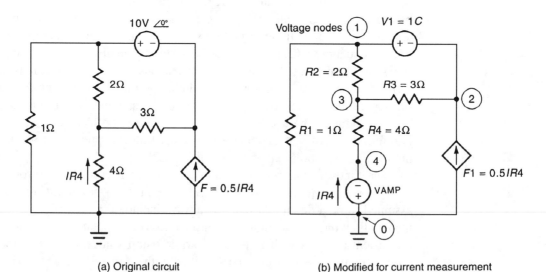

(a) Original circuit   (b) Modified for current measurement

Figure 6.15   Sample circuit for SPICE file "SAMPLE1"

The program for sample circuit, SAMPLE1.CIR, is shown in the following table:

SAMPLE1.CIR

Element	Nodes	Control	Value	Remarks
R1	0 1		1	Ohms
R2	1 3		2	Ohms
R3	3 2		3	Ohms
R4	3 4		4	ohms
V1	1 2	DC	10	volts
F1	0 2	VAMP	0.5	conductance
VAMP	0 4	DC	0	Zero Voltmeter current
.END				

The SPICE solution is given as

Node	Voltage	Currents (through voltmeters)
1	0.8955 V	V1 → −2.537 A
2	−9.104 V	VAMP → 0.597A
3	−2.388 V	
4	0.0 V	

The preceding sample problem and a somewhat more comprehensive one for an ac steady-state circuit (to follow shortly) are given only to jog the reader's memory on problem/solution methods. If such problems were given on the PE examination, then all of the symbols and control statements for a particular program would have to be given (which is normally not practical on an examination). In this author's opinion, only the concept is important.

For furthering one's concept of problem formulation and solution, a few more remarks about SPICE for ac circuits will be made.

The on-line control statement for an ac sinusoidal analysis is

```
.AC LIN NP FSTART FSTOP
```

Here, LIN means linear variation, NP is the number of points (*i.e.*, frequency points), FSTART and FSTOP are the starting and stopping frequencies in hertz (for a fixed frequency operation, they are the same). Another control statement is required for the output. For rectangular quantities, the difference between real (R) and imaginary (I) values must be indicated. For polar quantities, the magnitude (M) and the phase (P) are specified. For instance, for a desired voltage, one would specify $VM(n_1, n_2)$ for the voltage magnitude from node n_1 to n_2. The whole line might be

```
.PRINT AC VM(3,4) VP(3,4)
```

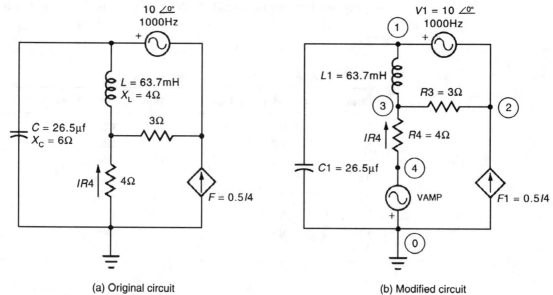

(a) Original circuit   (b) Modified circuit

Figure 6.16   An ac for SPICE file: "ACSAMPLE.CIR"

The resulting output would be the magnitude and the phase of the voltage between nodes 3 and 4 (with the polarity being positive at node 3 with respect to node 4). Consider the following RLC sample problem (see Fig. 6.16).

The program for the ac sample problem, ACSAMPLE.CIR, is shown in the following table (the voltage across and the current through the 4 Ω resistor):

ACSMPLE.CIR

Element	Nodes	Control	Value	Remarks
C1	0 1		26.5E-6	
L2	1 3		63.7E-3	
R3	3 2		3	
R4	3 4		4	
V1	1 2	AC	100	zero is the angle in
F1	0 2	VAMP	0.5	degrees transconductance
VAMP	0 4	AC	0	
.AC	LIN 1	1000	1000	
.PRINT	AC	VM(3 4)	VP(3 4)	IM(VAMP)
.END				

The computed answers are: VM = 3.51, VP = −127.5°, IM = 0.879. As a check, the voltage across the 4 Ω resistor, VM(3 4), is the current through the fictitious ammeter, VAMP. It is found to be 4 × IM(AMP) = 4 × 0.879 = 3.51 volts.

There are many other control statements for other operations, such as sensitivity (.SENS) to parameter variation, transfer function calculations (.TF), etc., that may be used in SPICE. These operations are beyond the scope of this review.

As another example, a suitable Spice input file for the circuit in Figure 6.17 would be

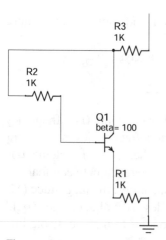

Figure 6.17   A BJT transistor

```
VS  1 0 DC 1
R1 1 2 DC 1
RI 2 0 500K
R2  2  3  50K
R3  3  4  100
EX 4  0  0 2  100k
.TF  VS  V(3)
.END
```

CONTROL SYSTEM SIMULATION—CSMP

The subject of simulation (of control systems) is very broad. There are many different programs available, including SPICE. However, for this section, only programs specifically tailored to control systems, such as CSMP, are discussed. These kinds of dynamic-system simulation languages fit into a rough standard as set forth by Simulation Council, Inc. They all allow direct entry of first-order differential equations and give automatic translation of routines into the particular language being used.

Although CSMP is an older language continuous simulation program, it is still one of the most popular and contains most of the elements of the newer programs (such as ACSL, CSSL, CC, DESIRE, MATRIXx, PCESP, TUTSIM, etc.). Generally these programs were developed for mainframe computers. However, most have smaller versions developed for the IBM PC and are well presented in applicable literature (see the Recommended References at the end of this chapter). Although a short review of the CSMP program is presented here, an excellent older text (Speckhart and Green) is available in many technical libraries for a more thorough study on the subject of simulation.

Using CSMP is more like using an analog computer rather than writing programs in a higher-level language. CSMP may be used without reference to FORTRAN but, for more advanced applications, a background in FORTRAN is helpful. As with all of these simulation packages, prepackaged algorithms for integration and many other operations are available (within the programs). For example, several different integration approximation algorithms are available on which best suits a particular problem. If no integrating method is specified, the program automatically defaults to a variable step type such as the Runge-Kutta-IV or the Milne method). The *A default routine is usually more than adequate for all but the more specialized type of problems (e.g., widely separated time constants, discrete time-delayed sample and hold functions, etc.). For sample and hold simulations, a fixed step algorithm should be used.

CSMP program statements usually fit into three categories. These categories are data statements, structure statements, and control statements. Generally, the structure of CSMP has three segments to it: the initial segments, the dynamic segments, and the terminal segments. Initial conditions and direct calculations (e.g., the area of a circle) may be placed in the initial segment. The dynamic segment should include anything involving system iterative computing (such as the describing differential equations). The terminal segment may include control statements, plotting algorithms or routines, output statements, and any timing statement. Arithmetic operations use the same symbols as in FORTRAN. Automatic sort routines are incorporated and may be assumed unless specifically called out with a nosort statement. Nosort would be used for special procedural subroutines such as branching and counting. To discuss all but the simplest concepts, it is necessary to know the version of CSMP that is available and to have the necessary manual defining the various block and integration routines.

A very simple example is given to illustrate the method. First, a few definitions and blocks are given in the following table:

Program Statement	Function	Block Diagram
`Y = INTGRL(IC,X)`	$y = \int x\,dt + IC,$	$X \rightarrow \boxed{\dfrac{1}{s}} \rightarrow Y$
`Y = REALPL(IC,TC,X)`	$TC\dot{y} + y = x,$	$X \rightarrow \boxed{\dfrac{1}{(TCs+1)}} \rightarrow Y$
`Y = STEP(T)`	$y = 0$; for $t<T$, $y = 1$; for $t>T$	(step graph at T)
`Y = LIMIT(N`$_1$`,N`$_2$`,X)`	$y = N_1$; for $x<N_1$; $y = N_2$; for $y>N_2$ $y = x$; for $N_1<x>N_2$	(limit graph N_1, N_2)

Assume one wishes to simulate the nonlinear block diagram shown in Figure 6.18.

In order to avoid differentiating for analog and digital simulation, it is convenient to rearrange the problem as shown in Figure 6.19.

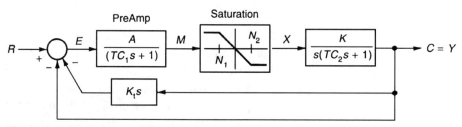

Figure 6.18   A nonlinear problem

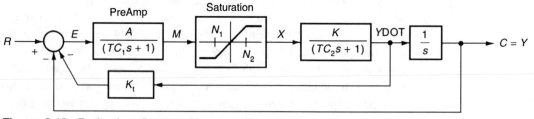

Figure 6.19   Revised nonlinear problem (see Fig. 6.18)

The corresponding CSMP simulation program is

```
TITLE PROGRAM TO SIMULATE FEEDBACK SYSTEM
  INITIAL
  CONSTANT A = 5.0, K = 2.0, TC₁ = 0.005, TC₂ = 0.5, N₁ =
- 1.0, . . .
      N₂ = 1.0, K_T = 0.025
  DYNAMIC
    E = STEP(0) - K * YDOT - Y
    M = REALPL(0, TC₁, E)
    X = LIMIT(N₁, N₂, M)
    YDOT = REALPL(0, TC₂, X)
    Y = INTGRL(0, YDOT)

*COMMENT  THIS COMPLETES THE DYNAMIC PORTION
  TERMINAL
    TIMER FINTIM = 2.5, PRDEL = 0.1, DELT = 0.01
    PRINT Y, E, M
    END
    STOP
*COMMENT  PROGRAM FINISHES IN 2.5 SEC., THE PRINTOUT IS EVERY
*COMMENT  PRDEL SECONDS, BUT THE MINIMUM STEP SIZE OF THE
*COMMENT  INTEGRATION INTERVAL IS TO BE DELT
*COMMENT  IF DEL IS NOT SPECIFIED, THE DEFAULT IS
*COMMENT  DEL = FINTIM/100
```

Because the integration method is not specified, the default is a variable step size Runge-Kutta method. If the system contained a sample and hold portion for z-transforms, a statement specifying a fixed step integration method should be inserted. Also, the DEL would have to be some sub-multiple of the output interval (the first step is 1/16th of PRDL or OUTDEL). As an example,

```
METHOD RKSFX

TIMER FINTIM = 2.0, DELT = 0.01, OUTDEL = 0.16

*COMMENT RKSFX IS A FOURTH ORDER RUNGE-KUTTA WITH FIXED
INTERVAL
```

To review for any of the simulation programs, one needs the appropriate manual. Two or three of the integration routines besides the various block calls could be covered in such a review.

This particular package (CSMP) was selected for review because of the great number of text references available. In studying for the PE examination, one should not focus on a particular language but rather should study the concepts underlying how languages are structured.

RECOMMENDED REFERENCES

Korn, *Interactive Dynamic System Simulation*, McGraw-Hill, 1989.

Nilsson and Riedel, *Introduction to PSpice for Electric Circuits*, 6th Ed., Prentice-Hall, 2002.

Shah, *Engineering Simulation: Tools and Applications Using the IBM PC Family*, Prentice-Hall, 1988.

Speckhart and Green, *A Guide to Using CSMP—The Continuous System Modeling Program*, Prentice-Hall, 1976.

Digital Logic and Systems

OUTLINE

This chapter will review three areas of study: (1) digital logic, (2) digital interfacing and buses, and (3) digital systems. Because this field has grown so fast and now covers almost every area in electrical design and instrumentation, review of this chapter is recommended even if just for notational purposes. (Readers who already have expertise in this area may choose to skip this chapter.) For those who have no background in digital logic, many books are available for additional study. Two very popular texts (used at the undergraduate level) were written by Bartee and by Roth; the author particularly recommends these. A somewhat more advanced text by Hill and Peterson is also especially well written and recommended.

DIGITAL LOGIC

Understanding the operation of most digital systems requires knowledge of digital logic and Boolean algebra. It will be assumed the reader is already familiar with number conversion from one base to another, such as binary to octal or to hexadecimal, and also comfortable with binary arithmetic operations.

While integrated circuits have taken over the tedium of individual gate design, it is best to start this review with individual gates and how they implement Boolean algebra operations. Unlike normal algebra, where a variable may represent almost any value, the variables in Boolean algebra must be either a 0 or a 1. Also, the "+" sign takes on the meaning of OR while the "·" means AND (however, the dot is frequently not shown). These kinds of operations may not necessarily be used in any order, but must follow certain rules. Also, in Boolean algebra, the operation of complementation (designated with a bar over the variable $\overline{A}$ or, for convenience, sometimes written as A') must be understood; thus if A = 1, then A' = 0 (read as "not A"). These operations will lead directly to other gating such as NAND and NOR gates. Other symbols may also represent these connectives. These connectives and other definitions will be shown in the following two tables.

Term	Symbols
AND	$\wedge, \cdot, \cap$
(Inclusive) OR	$\vee, +, \cup$
Exclusive OR	$\oplus, \forall$
NOT	$—, \sim, ', *$
NOT and (NAND)	$\uparrow —$
NOT or (NOR)	$\downarrow —$
If A is true, then B is True	$\supset$

Term	Definition	Synonym
Literal	A variable or its complement (A, $\overline{A}$, B, $\overline{B}$)	
Product term	A series of literals related by AND (A $\overline{B}$ D)	Conjunction
Sum term	A series of literals related by OR ($\overline{A}$ + B + $\overline{D}$)	Disjunction
Normal term	A product or sum term in which no variable appears more than once	

These operations are given as combinational logic, while their respective symbols and circuits, as used in logic design, will follow shortly. First, however, some of the rules, laws, theorems, and postulates (known as Huntington's postulates) of Boolean algebra are presented.

I.   There exists a set of K objects or elements, subject to an equivalence relation, denoted "=", which satisfies the principle of substitution. By substitution is meant if $a = b$, then a may be substituted for b in any expression involving b without affecting the validity of the expression.

IIa.   A rule of combination "+" is defined such that $a + b$ is in K whenever a or b is in K.

IIb.   A rule of combination "·" is defined such that $a \cdot b$ (abbreviated ab) is in K whenever both a and b are in K.

IIIa.   There exists an element 0 in K such that, for every a in K, $a + 0 = a$.

IIIb.   There exists an element 1 in K such that, for every a in K, $a \cdot 1 = a$.

IVa. $a + b = b + a$ (commutative laws)

IVb. $a \cdot b = b \cdot a$ (commutative laws)

Va. $a + (b \cdot c) = (a + b) \cdot (a + c)$ (distributive laws)

Vb. $a \cdot (b + c) = (a \cdot b) + (a \cdot c)$ (distributive laws)

VI.   For every element a in K there exists an element a such that

$$a \cdot \bar{a} = 0 \quad \text{and} \quad a + \bar{a} = 1$$

VII. There are at least two elements X and Y in K such that $X \neq Y$.

Duality

Every theorem that can be proved for Boolean algebra has a dual that is also true, as shown below:

$$a + (b \cdot c) = (a + b) \cdot (a + c)$$
$$a \cdot (b + c) = (a \cdot b) + (a \cdot c)$$

DeMorgan's Law (Theorem)

For every pair of elements a and b in K,

$$\overline{(a \cdot b)} = \bar{a} + \bar{b}$$
$$\overline{(a + b)} = \bar{a} \cdot \bar{b}$$

Associative Laws

For any three elements, a, b, and c in K,

$$a + (b + c) = (a + b) + c$$

and

$$a \cdot (b \cdot c) = (a \cdot b) \cdot c$$

Gates

Combinational logic is made up of groups of AND and OR gates, with their many variations. These have been implemented over the years as diode logic, resistor-transistor logic (RTL), diode-transistor logic (DTL), direct-coupled transistor logic (DCTL), transistor-transistor logic (TTL), emitter-coupled logic (ECL), and others. There is negative logic ($V^- = 1$, $V^+ = 0$), which is rarely used today, and positive logic ($V^- = 0$, $V^+ = 1$). The voltage swing away from ground can be either positive or negative. Transistors used can be either **PNP** or **NPN.** No matter how gates are implemented, hardware-wise, their inputs and outputs can assume only one of two states at any one time. Circuits for various gate hardware mechanizations and their truth tables are shown in Figures 7.1 through 7.5. These different kinds of gates are used in combinational logic circuits that are derived directly from problem statements.

The gates use transistors and diodes. However, the transistors use only a very small current to control themselves (for large 8's) but can carry a large current in the ON mode of operation. For these ON and OFF operations, the transistors

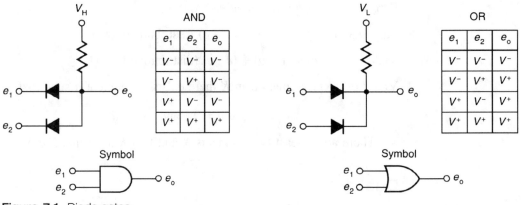

Figure 7.1   Diode gates

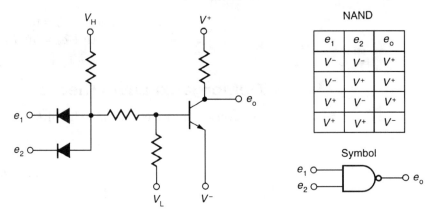

Figure 7.2   DTL gate

operate in the nonlinear region, being either in the saturation mode or in the cutoff region. In transistor switching circuits, a very good approximation for determining logic functions is to consider an ON transistor as a LOW output; here, a three-terminal short-circuit device as in Figure 7.6a brings the collector voltage to near zero (a saturated condition). On the other hand, an OFF transistor has a three-terminal open-circuit device with a HIGH at the collector terminal. (Of course, the collector resistor along with external loading determines the voltage at the collector terminal.) The saturated condition must be verified by checking the resistor values and the V_{cc} supply against the minimum current gain specified for

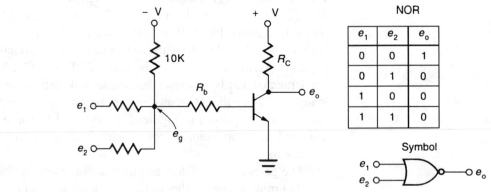

Figure 7.3   Typical TRL gate

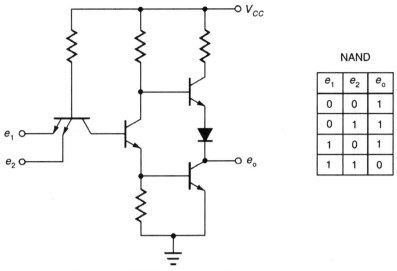

Figure 7.4 TTL gates (7400 series)-NAND

NAND

e_1	e_2	e_o
0	0	1
0	1	1
1	0	1
1	1	0

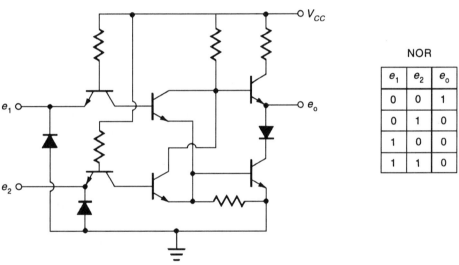

Figure 7.5 TTL gate (7400 series)

NOR

e_1	e_2	e_o
0	0	1
0	1	0
1	0	0
1	1	0

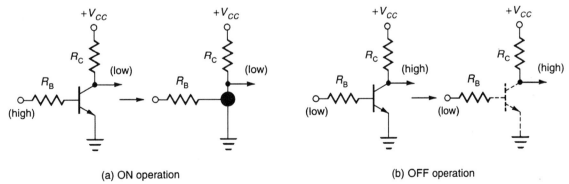

(a) ON operation

(b) OFF operation

Figure 7.6 Nonlinear operation for a grounded emitter transistor

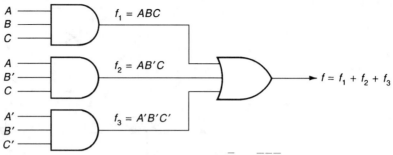

Figure 7.7 Implementation of $f = ABC + A\overline{B}C + \overline{A}\,\overline{B}\,\overline{C}$

the transistor. If the I_c/I_b ratio is smaller than the minimum h_{fe} specified, then saturation can be assumed and the logic levels may be easily determined.

As an example, assume it is desired to design a burglar alarm using digital gates with three (A, B, and C) ON-OFF type sensors; furthermore some sensors are connected such that if they are in the OFF condition they may also cause an alarm to ring if other stipulated conditions are met. Assume the design yields the following equation (where the function $f(A,B,C)$ if 1, causes the alarm to ring):

$$f(A,B,C) = f = ABC + A\overline{B}C + \overline{A}\,\overline{B}\,\overline{C}$$

The alarm implementation could be given as shown in Figure 7.7.

Karnaugh Maps

Using Boolean algebra the expression for f (as shown in Fig. 7.7) may be shown to be equivalent to (recall, the bar over a letter means the same as an apostrophe)

$$f(A,B,C) = AC(B + \overline{B}) + \overline{A}\,\overline{B}\,\overline{C} = AC(1) + \overline{A}\,\overline{B}\,\overline{C} = AC + \overline{A}\,\overline{B}\,\overline{C}.$$

This reduced expression for f eliminates the need for one AND gate. For a large system this reduction (minimization) technique is greatly simplified with the use of a Karnaugh map and its corresponding truth table; the truth table for the same function, with the assigned min-term numbers (repeated on the Karnaugh map), is shown (see Fig. 7.8). In general, from a problem statement, a truth table is formed from the natural binary code.

As another example, assume that from three ON/OFF switches, A, B, and C, it is required to have an output signal that is ON (high) if switches A and C are ON, or if B and C are ON but only if the digital number that represents the signals from A, B, and C is odd. Also, if these three signals represents the digital number

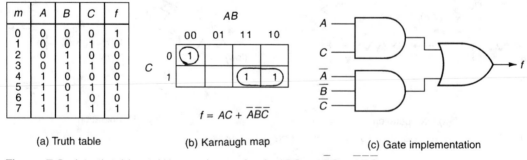

| (a) Truth table | (b) Karnaugh map | (c) Gate implementation |

Figure 7.8 A truth table and Karnaugh map for $f = ABC + A\overline{B}C + \overline{A}\,\overline{B}\,\overline{C}$

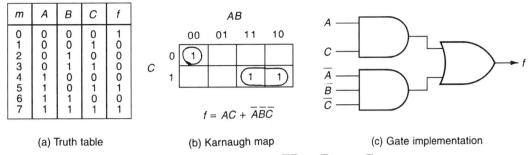

m	A	B	C	f
0	0	0	0	1
1	0	0	1	0
2	0	1	0	0
3	0	1	1	0
4	1	0	0	0
5	1	0	1	1
6	1	1	0	0
7	1	1	1	1

$$f = AC + \overline{A}\,\overline{B}\,\overline{C}$$

(a) Truth table   (b) Karnaugh map   (c) Gate implementation

Figure 7.9 A truth table and Karnaugh map for $f = \overline{A}\,\overline{B}\,\overline{C} + A\overline{B}C + ABC$

four, we don't care whether the output is ON or OFF (the "don't care" term may be represented by "dc" or sometimes, an "X"). The truth table is given and the minimal sum of products expression is found from the grouping in the Karnaugh map as shown in Figure 7.9.

Decoders and Multiplexers

For decoding a group of signals, recall that a standard decoder may also be used instead of the above network to give the desired output signal as shown in Figure 7.10. (As a help in starting your review, you should first check the internal circuit of a typical 1 out of 8 decoder; or, even better, derive the circuit from a truth table.)

Recall the function of a multiplexer (MUX); it selects a particular line from several lines and passes the information directly to its output. Consider the circuit equivalents in Figure 7.11.

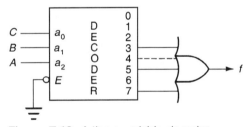

Figure 7.10 A three-variable decoder

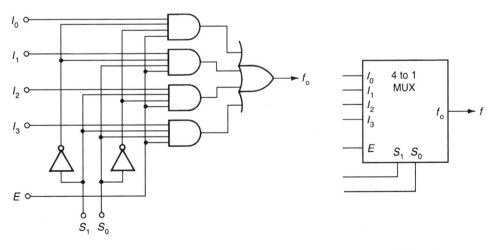

(a) Circuitry   (b) Symbol

Figure 7.11 A 4-to-1 multiplexer

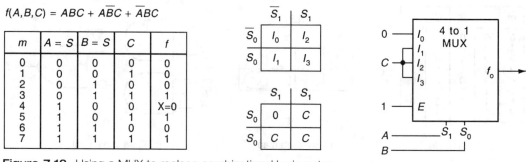

$f(A,B,C) = ABC + A\bar{B}C + \bar{A}BC$

m	A = S	B = S	C	f
0	0	0	0	0
1	0	0	1	0
2	0	1	0	0
3	0	1	1	1
4	1	0	0	X=0
5	1	0	1	1
6	1	1	0	0
7	1	1	1	1

	$\bar{S_1}$	S_1
$\bar{S_0}$	I_0	I_2
S_0	I_1	I_3

	$\bar{S_1}$	S_1
$\bar{S_0}$	0	C
S_0	C	C

Figure 7.12 Using a MUX to replace combinational logic gates

Now consider the use of the MUX as decoder by reference to a truth table and a Karnaugh map. Assign as many variables as possible to the select lines; one then could assign the other variables through logic combinations to the MUX inputs. Referring to the example shown in Figure 7.11, from Figure 7.12 it is clear that by using both normal inputs ($I_{0,1,2,3}$) and select inputs ($S_{1,2}$) it is possible to use a MUX as a decoder.

At this point in the discussion of decoders and multiplexers (and other devices yet to be covered), it is important to understand a modified notation that accounts for lows and highs in a circuit and in some cases negative logic (meaning that a 1 corresponds to a low and a 0 means high—rarely used today). However, a mixed logic notation, using blocks rather than the traditional gate symbols, that allows one to define the logic level at any point in a circuit is sometimes used (and, in fact, there is a ANSI/IEEE, STD 91, 1984 that defines this notation). Consider first some alternatives for standard AND, OR, NAND, and NOR gates as shown in Figure 7.13. (It is suggested that the reader prove these relationships by using DeMorgan's Theorem.) These alternate symbols are also defined as rectangular blocks with the notation as shown in Figure 7.14.

It should be noted that the small circle, NOTing or inverting symbol, has been replaced with a half arrow—meaning that a logic 1 (internal to the device) yields an L at the output (the significance of the small circle and the half arrow is virtually the same for positive logic—but has different meaning for negative logic).

With the preceding remarks for alternate gating symbols, the application of the ANSI/IEEE standard can now better be explained with regard to a dependency notation. This dependency notation is used to show the relationship between the input and the output. As an example, the traditional Enable (E) line is replaced with (EN) if the output is enabled for a particular input. A simple decoder (referred

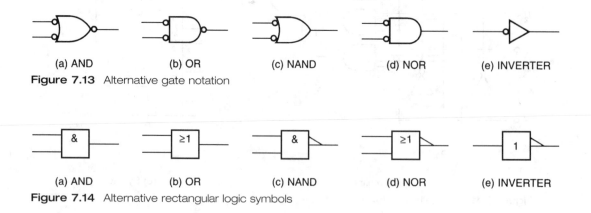

(a) AND	(b) OR	(c) NAND	(d) NOR	(e) INVERTER

Figure 7.13 Alternative gate notation

(a) AND	(b) OR	(c) NAND	(d) NOR	(e) INVERTER

Figure 7.14 Alternative rectangular logic symbols

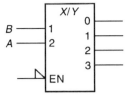

Figure 7.15 A 2-to-4 line decoder

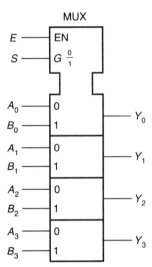

Figure 7.16 A quadruple 2-to-1 MUX using NASI/IEEE standards

to as a member of the "coder" family) would now be labeled as in Figure 7.15; here, the numbers on the input refer to a binary weighted line of $2^0 = 1$, $2^1 = 2$, and $2^2 = 4$.

The concept is that if input lines 1 and 2 are both active then the weighted input is $1 + 2 = 3$, which results as the active output line number. Another feature of the ANSI/IEEE standard is that of dependency; for an AND gate dependency, the letter G (followed by a number) defines an input (or output) number ANDed with that G input (for an OR gate dependency, the letter V is used, etc.). For instance, if a block diagram has several inputs with, say, two G's (*i.e.*, $G1$ and $G2$) and several other inputs with, say 0 through 3, this immediately tells us that output is related to input by an AND gate; thus for $G2$ the output is related to two inputs, line $G2$ and line 2 by means of AND function ($G2$ might be a select line while 2 might address line 2). On the other hand, if an input has a notation and is AND dependent on, say 8 input address lines (*i.e.*, 0 through 7), all address lines, then a shorthand notation would be $G\frac{0}{7}$.

Also, if there is control block (or a group of control lines) in a block diagram, the diagram would separate with the other functions of the block with a notch. As an example, consider a quadruple two-input MUX (see Fig. 7.16). The enable line and the select line control the whole function of the MUX as is shown in the upper block.

While this ANSI/IEEE standard is quite insightful if fully understood, it also can be confusing and it is not widely used. In this author's opinion any question on the PE examination probably will use the straightforward standard symbols. However, if one wishes to pursue this subject without having to go through the full ANSI/IEEE standard, then a number of texts go into more detailed explanation.* The presentation of the rest of material in this chapter will continue to use the normal logic symbols for clarity.

Read Only Memories

Before flip-flops and active memory are discussed, a Read Only Memory (ROM) will be reviewed at this point. It should be recalled that binary values may be stored in a matrix form and could be represented by either a voltage near 5 volts (1) or low voltage (0) and this voltage may be obtained from the voltage drop across the resistor. Consider the following oversimplified circuit where the current is directed by diodes; missing diodes mean no current flows. (Note that the crossing wires are non-touching.) By selection of the address, or rotary switch number, the particular horizontal wire is addressed; then current may or may not flow in the vertical data, output wires. This output data, $(d_o...d_n)$, appears on the data lines when the switch is at a certain position (particular address) as presented in Figure 7.17.

Of course, the mechanical switch is replaced with a decoder (Figure 7.18), which is normally included in the ROM package. The address is in the form of the binary code such that if one knew the number of address lines (al), the number of horizontal lines is given by 2^{al}. Also note the number of data output lines is independent of the number of address lines. Again, refer to the previous example; we may use addresses as the signal lines and connections by the diodes will yield the desired output.

*See Mano, *Computer Engineering Hardware Design,* Prentice Hall, 1988, Pages 108–115 or Roth, *Fundamentals of Logic Design,* West Publishing Company, 3rd ed., 1985, Appendix B.

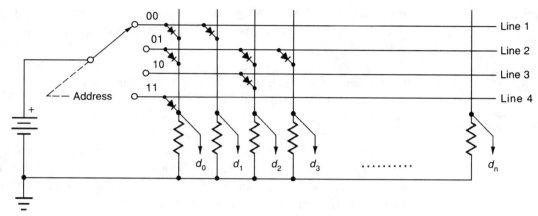

Figure 7.17   A fixed ROM memory

Of course, there are a number of different kinds of so-called fixed memories, such as erasable programmable read only memory (EPROM), flash memory, etc.; some of these will be addressed later.

Flip-Flops

One of the most important elements of a digital system is the ability to store information in an active manner. The common hardware storage device is the flip-flop, (f-f), or a bistable multivibrator. Before discussing the internal operation of these devices, the overall description of the device family follows.

(a) *D.*   A flip-flop whose output is a function of the input that appeared just prior to the clock pulse.

(b) *J-K.*   A flip-flop having two inputs, *J* and *K*. At the application of a clock pulse, 1 on the *J* input will set the output to 1; 1 on the *K* input will reset the output to 0; 1 on both inputs will cause the output to change state (toggle); O on both inputs results in no change in the output state.

(c) *R-S.*   A flip-flop having two inputs, *R* and *S*. Operation is the same as the *J-K* flip-flop except that 1 on both inputs is illegal.

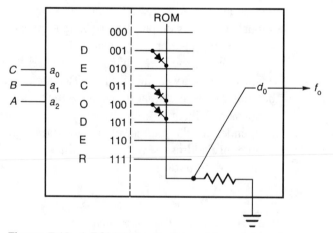

Normally many other output data lines are available for stored data; this is an over-simplified example.

Figure 7.18   A ROM that includes an internal decoder

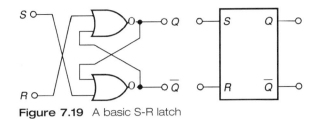

Figure 7.19 A basic S-R latch

(d) *R-S-T.*   A flip-flop having three inputs, *R*, *S*, and *T*. The *R* and *S* inputs produce outputs as described for the *R-S* flip-flop; The *T* input causes the flip-flop to toggle.

(e) *T.*   A flip-flop having only one input. A pulse appearing on the input causes it to toggle.

These various kinds of f-f devices may be used as a latch or may be edge triggered by an external signal. The basic circuit may be made up from a pair of NAND or NOR gates. Consider the NOR pair and its flip-flop representation. Call one input SET (S), the other RESET (R), one output Q, and the other $\bar{Q}$; this f-f is called a latch (see Fig. 7.19).

If, on the other hand, the signals S and R are withheld by a pair of AND gates, the f-f will stay in whatever state it happens to be in unless the signal C is also on. A further step would be to allow the signal C to have an effect only when the leading edge of a pulse is applied (think of this as allowing the wave front through the capacitor and the function of the resistor is the discharge for the capacitor). The leading edge of the C signal will cause any signals on S or R to get through; this C signal is thought of as the controlling clock signal (*i.e.*, the leading edge of a pulse) as shown in Figure 7.20.

The most popular versions of these edge triggered f-f's are the "T", "D", "J-K", and the master-slave units. Recall that the time delay through the f-f allows us to use these devices in synchronous circuits (*i.e.*, all driven by a common clock). Also recall that the truth table for the *S-R* and *J-K* f-f's are the same EXCEPT the inputs on the *S-R* f-f should never have 1's at the same time, while the *J-K* may both have 1's on their inputs (Figure 7.21).

As an example consider the configuration shown in Figure 7.22; assume all Q's are cleared to start (i.e., set to zero).

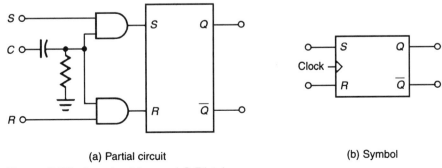

(a) Partial circuit   (b) Symbol

Figure 7.20 An edge triggered *S-R* latch

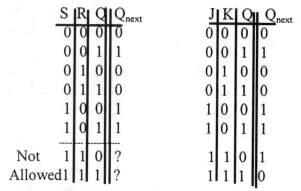

S	R	Q	Q_next
0	0	0	0
0	0	1	1
0	1	0	0
0	1	1	0
1	0	0	1
1	0	1	1
1	1	0	?
1	1	1	?

Not Allowed (rows 1 1 0 ? and 1 1 1 ?)

J	K	Q	Q_next
0	0	0	0
0	0	1	1
0	1	0	0
0	1	1	0
1	0	0	1
1	0	1	1
1	1	0	1
1	1	1	0

Figure 7.21 Truth tables for inputs on the *S-R* and *J-K* flip-flops

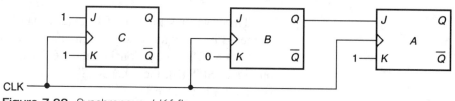

Figure 7.22 Synchronous *J-K* f-f's

Timing Diagrams

Figure 7.23 shows the timing diagram for Figure 7.22. It should be noted that if one neglected the time delay through the f-f's, the timing diagram would not take into account the correct signals "ready and waiting" on the respective inputs when the steep wave front of the clock signal arrives. Here, one assumes the clock signal acts of the f-f's before the output of the previous f-f has changed.

Most commercial flip-flops have two other inputs for either setting Q initially to a 1 (called a PRESET), or clearing Q to 0 (called CLEAR or sometimes RESET). This is because clock signals are frequently left connected, but with these other signals the device is held at the desired value until ready. As an example, consider the following sample design for an UP-DOWN counter.

For an up count signal (enable) on the UP-DOWN counter of Figure 7.24, on every clock pulse (after the DC RESET line is Released) the output X's (with X1 being the least significant bit) will read 000, 001, 010, 011, 100, ... 111, then start over again. The reader is encouraged to trace through the circuit and make his

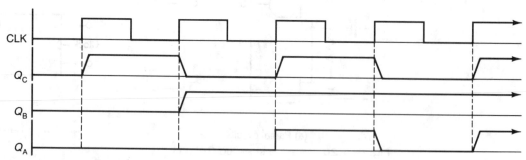

Figure 7.23 Timing diagram for Figure 7.22

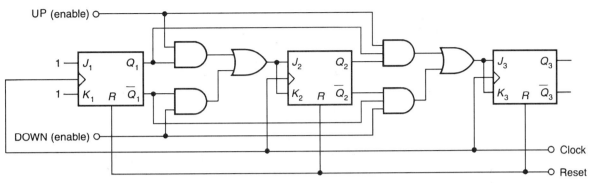

Figure 7.24 An UP-DOWN synchronous counter

own timing diagram to confirm its operation, then to repeat the operation for a down count.

Sequential Counter

Consider the design of a sequential "odd ball" counter that has a sequence of, say, 0,1,3,2, ... repeat. Because four different values are represented, only two flip-flops are needed (*i.e.,* $2_n = 4$, then $n = 2$). Assume only one *J-K* type and one *R-S* type f-f is available; a state table may be developed along with a state diagram to help one formulate the necessary logic. Designate one f-f as the most significant bit (MSB), say the *J-K* type f-f, and the other f-f as the least significant bit (LSB). The state table may be completed from the truth table of these two types of f-f (Figure 7.25), where

- *PS* is the present state

- *NS* is the next desired state

- the *J-K* f-f is designated as A

- the *S-R* f-f is designated as B.

Karnaugh maps may be used for each input of each f-f to form the circuit logic; then a timing table is sketched to check for the correct operation (see Fig. 7.26).

For practice, try designing a modulo 5 counter (*i.e.,* one that counts from 0 to 5 then repeats) using all *J-K* f-fs. The final solution is given in Figure 7.27.

m	PS		NS		J-K		S-R	
	MSB	LSB	MSB	LSB	J	K	S	R
0	0	0	0	1	0	X	1	0
1	0	1	1	1	1	X	X	0
3	1	1	1	0	X	0	0	1
2	1	0	0	0	X	1	0	X
0	REPEAT ------------------							

(a) Truth table

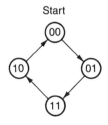

Start

(b) Sequence graph

Figure 7.25 An "odd ball" sequence counter

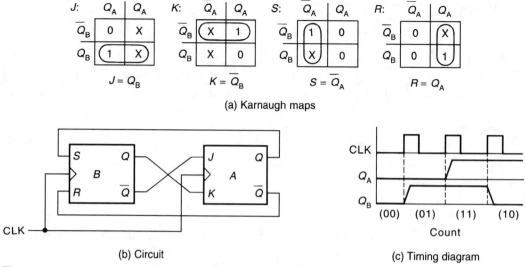

(a) Karnaugh maps

(b) Circuit

(c) Timing diagram

Figure 7.26 Solution and implementation of the "odd ball counter"

Memory—RAM

Random Access Memory banks get larger, faster, and more complex every year. However, at this point, only the basis concept of operation, starting with the single cell, then progressing through address decoding to access larger groups of cells, will be reviewed. First consider each cell, C, as an individual flip-flop: Each cell needs to be controlled so that a binary quantity may be transferred into, or out of, the cell on a Read/Write command when the cell is selected. Such a controlling circuit could be given as follows in Figure 7.28 (in this case, without a timing signal).

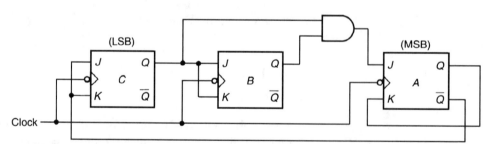

Figure 7.27 A modulo 5 counter

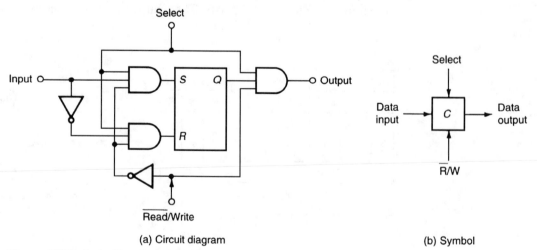

(a) Circuit diagram

(b) Symbol

Figure 7.28 A single cell of a random access memory

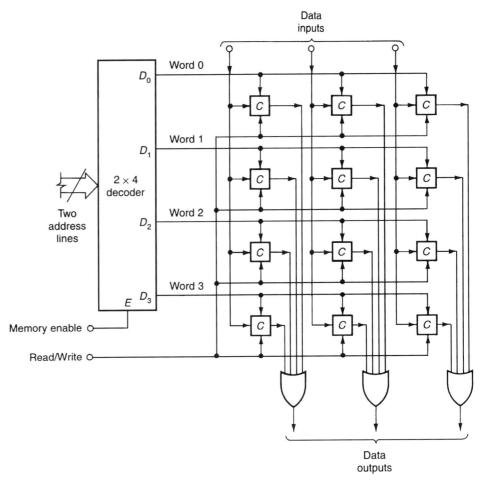

Figure 7.29 An oversimplified 4 × 3 RAM

These cells are normally contained as an integrated circuit with an included decoder such that each word (of m words) with n data bits may be selected for reading or writing on an individual basis. The select line (called a word select line) enables data to be written to or read from that particular word (independent of the number of data bits). An oversimplified memory of a 4 word (0 through 3) with only 3 data bits ($m \times n = 4 \times 3$) of data is shown below (see Fig. 7.29).

Other control lines, such as (1) chip select, CS (in addition to the address word select), and (2) output enable lines, OE (in addition to the read line), may also be needed. These control lines are shown in Figure 7.30. A word of caution is needed when using the "enable" term. Actually the output of a gate should either be a 0 or a 1; however, for certain conditions (especially when the output goes to a bus), one would like to disconnect the output or leave it floating. To achieve this floating effect, the gates can be made to offer a tristate mode of operation (that is, the output may either be a 1 or 0 or floating). For this tristate operation, the gate operates normally when the enable signal is active, and is effectively disconnected from the output line when the enable line is not activated.

Consider a typical 4096 bit chip made up of 1024 (m-words) with 4 bits of data. The decoder for the word lines must have 10 address lines ($2^{10} = 1024$ or a 1 K × 4 RAM). Frequently it is desired to have 8 data bits; to use two 4096 bit

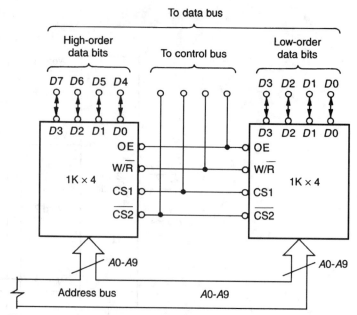

Figure 7.30 A 1K × 8 bit RAM using 1K × 4 bit RAM chips

chips to make a 1024×8 RAM, one merely makes use of the memory enable line on each chip to accomplish the desired 8-data bit path (see Fig. 7.30).

Because the data lines in Figure 7.30 operate in both the IN and OUT mode and can go to the same bus, the output lines should actually have a tristate gate (enable) so as not to load the data lines when in the input mode.

Memory—PLA & PALs

Programmable Logic Arrays are useful for replacing a complex logic circuit that may include extensive use of decoders. These devices are, in one sense, similar to ROMs; in fact, if the previously described ROMs were considered to be "sum ROMs" or OR gates, then another kind could be called a "product ROM" or AND gates. Rather than using just diodes as in the configuration of Figure 7.17 for ROMs, instead, use fused type links in a matrix (with the horizontal lines considered similar to AND gates, and the vertical lines as OR gates) as shown in Figure 7.31. Here, the lines cross each other but don't touch; the connections are made either by the links open ("blown" fuse links) or a connection through the fuse itself. Consider a previously programmed PLA, programmed for

$$f(A,B,C) = AB\overline{C} + \overline{A}\,\overline{B}.$$

Here, the blown links are shown as an open circuit, while an unblown fuse is shown as a dot (see Fig. 7.32).

A PAL (Programmable Array logic) is a special case of the PLA but is easier to program; the AND function is programmable but the OR function is not. However, the PAL, because of the limitation for the OR gates, usually requires the function be simplified as much as possible because the AND functions may not be shared with similar ones to the OR gate(s).

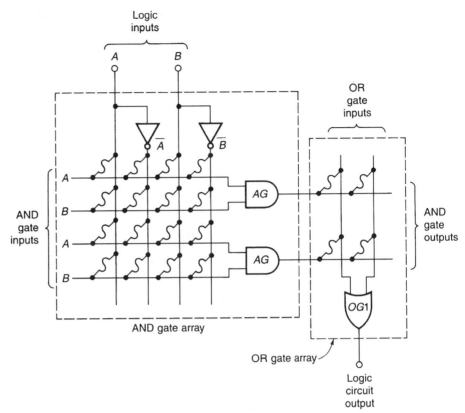

Figure 7.31 A simple two-input, single-output function, PLA (before being programmed)

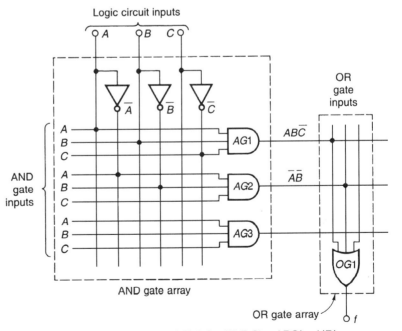

Figure 7.32 A preprogrammed PLA for f(A,B,C) = ABC' + A'B'

As an example of using a PLA to implement a certain design specification, assume the following design requirements: Two functions,

$$f_1 = ABC + \overline{A}\,\overline{B}C, \quad \text{and} \quad f_2 = \overline{A}\,\overline{B}C + AB\overline{C} + A\overline{B}C$$

are to be combined to form an output under a given timing order. The specified timing is such that upon initiating a START signal, the following conditions are to be met:

$$f_x = f_1, \quad \text{for} \quad 0 < t < 10 \text{ ms},$$
$$f_x = f_2, \quad \text{for} \quad 10 < t < 20 \text{ ms},$$
$$f_x = f_1 + f_2, \quad \text{for} \quad 20 < t < 30 \text{ ms},$$
$$f_x = 0, \quad \text{for} \quad 30 \ t > 30 \text{ ms (RESET condition)}.$$

The timing circuit is to be RESET, ready for the next START signal, after any 30 ms sequence. To implement this design, one could design a timing signal generator that might drive an f_1 and f_2 with enable control signals (see Fig. 7.33) and program a PLA for creating the two output signals. (The author urges the reader to do the actual design of the timing generator based on information presented in the last section.) For this problem, the PLA network would need at least 5 AND gate outputs, however, it should be noted that the two functions have common elements of A'B'C in f_1 and f_2 that may be used for both OR gates of the PLA. One such configuration (although not minimized) is as shown in Figure 7.34. In general, when using PLAs to implement a design, it is best to simplify the logic functions with a Karnaugh first to reduce the number of product terms because a PLA is limited in the number of gates. These PLA logic designs may be customized by the vendor or may be done in the field by using another type called FPLA (field programmable logic arrays); however, some FPLAs require special hardware programmer units.

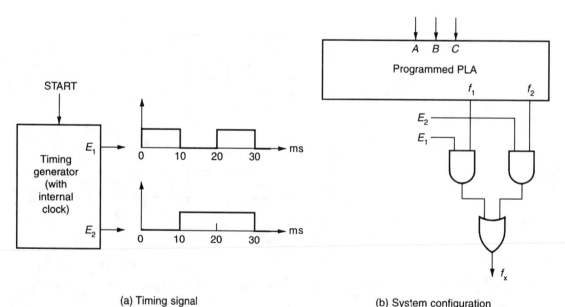

(a) Timing signal

(b) System configuration

Figure 7.33 Example problem block design

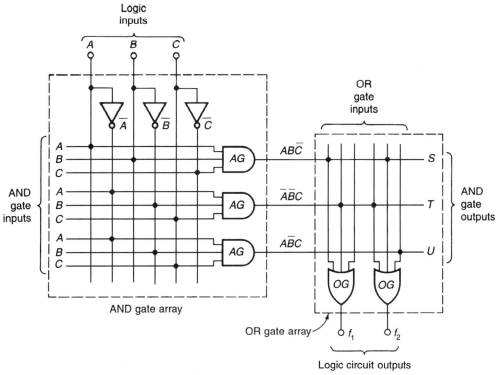

Figure 7.34 A PLA for $f_1 = A\overline{B}\overline{C} + \overline{A}\overline{B}C$ and $f_2 = \overline{A}\overline{B}C + AB\overline{C} + A\overline{B}C$

BUS INTERCONNECTIONS

A bus is normally considered as a common collection of wires that are interconnected to a group of devices that are timed-shared on these wires. The signal lines (or perhaps separate buses) frequently break into three groups: (1) command and control such as READ, WRITE, STATUS, etc.; (2) a data bus for a parallel system; and (3) address lines, usually for memories. Devices connected to the bus may take signals from the bus or furnish information to the bus; if anything connects to a bus, its input signals to the bus (with a low impedance) must be able to be "disconnected" when other signals are present. When the output is either a ONE or a ZERO, the gate is enabled, or if not enabled, then it is "floating." Consider a 4-wire bus; only one device at a time can WRITE (or talk) to the bus, but several devices may READ (or listen) to any signals on the bus (see Fig. 7.35). The READ

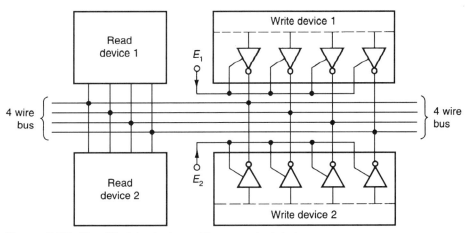

Figure 7.35 Simplified 4-wire bus with connected devices

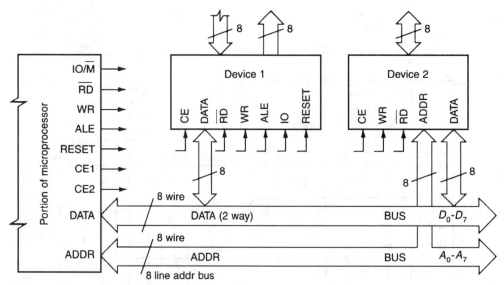

Figure 7.36   A simplified drawing of a bus connected system

devices almost always have enable control signals but they are not absolutely required for successful operation of the bus itself. Of course, connected devices, such as data paths, may have dual directions for READing and WRITing. Timing of the enable signals is crucial, such that if a device is enabled, any other write device must be left floating. A bus conflict will exist if timing signals are not very carefully controlled; as an example, if the output of two devices exposes their tri-state drivers (while enabled) to a conflicting voltage level, the resulting current may damage the gates.

For complex systems, the individual bus wires are not shown but are represented by a double line (or it could be one heavy line) with a slash mark indicating the number of wires and usually the wire identification numbers. As an example, consider a portion of a microprocessor connected to two peripherals (see Fig. 7.36).

For device #1, assume it has two ports, one for inputting data, the other for outputting data; for device #2 the port is two-way for inputting and outputting data (or for addressing certain specialized operations). Careful timing is required here for the CE's and the $\overline{\text{READ}}$/WRITE signals. It should be emphasized that Figure 7.36 is an idealized simplified microprocessor. (In actual practice, the chip enable lines 1 and 2 are normally part of the address bus.) The bus master here is the microprocessor and the two connected devices are called the slaves (when active). The convention for $\overline{\text{READ}}$ or WRITE is relative to the master (microprocessor) such that if $\overline{\text{READ}}$ is active, it takes information from the data bus. Shown in Figure 7.37 is a timing diagram for communicating to a particular device; it must be carefully related to a certain number of clock pulses. The sequence of operation for a READ cycle (from a device) is as follows (shown without clock pulses or handshaking):

(a) the ADDR lines must first be activated, then

(b) the CS line becomes high, then

(c) the DATA bus (which has been floating—high-Z state) is now in the active state, then

(d) the valid DATA is now presented to the microprocessor, then

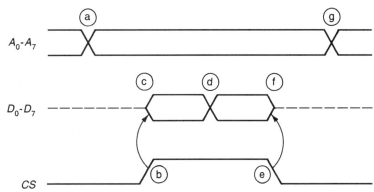

Figure 7.37 A simplified timing diagram of a READ cycle

(e) the device CS line goes low, which then causes

(f) the selected device to go back to floating, and, finally

(g) the cycle is complete and the microprocessor is ready for the next cycle.

To WRITE data, for example (assuming synchronous data transfer), the microprocessor makes the WRITE output line a zero, then the device must have time to READ its data onto the bus before the WRITE lines goes to a one. Of course, the address of the data bus should be active just before, during, and shortly after the data is being read.

IEEE488/GPIB and IEEE696 Standard Buses

This General Purpose Interface Bus (or as defined by the IEEE, 488 standard) was originally designed (by Hewlett-Packard) for interfacing programmable instrumentation and control systems. The original specifications were for an 8-bit system operating up to 1 Mhz for bus distances up to 20 meters (or for several devices at shorter distances). Later, a newer and larger bus, the IEEE-696 standard was developed that had a 24-bit address bus with a 16-bit data path and other desired features. Many different manufacturers have established their own standard buses, especially for the computer industry for their own application. For instance, IBM's PC original 8-bit bus, ISA bus (Industry Standard Architecture) became a standard. This ISA bus gave way to the 16-bit bus that had the original 8-bit configuration with card extensions (so as to be backward compatible) to the higher bit bus with faster speeds. Then came the EISA (Extended ISA) bus for 32-bit operation with speeds up to 10 MHz; however, with the addition of a Local Bus, the data could be transferred directly from peripheral to peripheral without having to pass through the CPU, thus allowing speeds to 66 MHz and above. For a more detailed (including pin numbers and slot sizes) explanation of the various buses, the reader is referred to "The Complete PC Expansion Bus Guide."

For purposes of explanation, the GPIB/IEEE-488 will be used in this discussion. This standard defines signal protocol as well as actual pin connections. The bus is a 24-pin standard that comprises 8-bit data lines, 5 lines for bus activity control, 3 for handshaking (or data control), and other lines for grounding. All 8 control lines use negative-true logic (*i.e.,* a low is equivalent to logic 1—not very common). Several terms used in this standard need to be defined as follows:

DAV (Data Valid)—Indicates the validity of data on the data bus (handshaking), a low (or a logic 1)

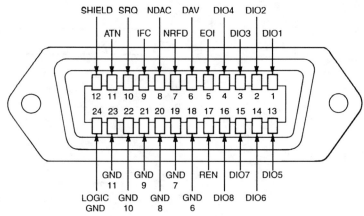

Figure 7.38 The IEEE-488 bus assignment

NDAC (Not* Data Accepted)—Indicates that data has been accepted by the listener (part of handshaking)

NRFD (Not* Ready For DATA)—Indicates the readiness of the data bus to accept data (part of handshaking)

ATN (Attention)—Indicates (when low) that data lines are used for commands or addresses; when high (logic 0), data is sent on this line

REN (Remote Enable)—Select whether remote or local control

IFC (Interface Clear)—Sets device to a known state

SRQ (Service Request)—Indicates device connected to bus needs attention

EOI (End Or Identify)—Indicates an ending of a sequence of events, or, if being polled, it identifies the device.

The bus coding and pin numbers are shown in Figure 7.38.

The bus may be grouped into three areas: (1) eight data I/O lines used for transfer of data between talkers and listeners (and sometimes used for transfer of commands between controller and devices), (2) five lines for control and management, and (3) three handshaking lines.

Handshaking

When the talker module wants to place information on the data lines, it first checks to see if all devices are set to listen by sending a low (logic 1) signal on the NRFD line. If the device(s) are ready, the NRFD line goes high, indicating to the talker to place data on the data bus (its DAV line then goes low). The control module then checks to see if the listener(s) have received the information by sampling the NDAC line (this line stays low until all listener devices have accepted the data; when all data transfers are complete, the NDAC line goes high; then the talker's DAV line goes high to indicate an acknowledgment. Of course, before handshaking can take place, a controller must designate which connected device to use; this is done by putting an address on the line corresponding to the particular device. A typical

---

*The "Not" notation implies negation is active.

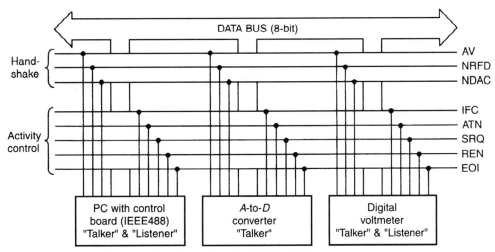

Figure 7.39 A typical configuration using the IEEE-488 bus

configuration might be where a PC-type computer might act as a controller (assuming it has a IEEE-488 interface card in it) and it is connected to a remote digital voltmeter and also an analog-to-digital converter as shown in Figure 7.39.

The timing diagram for handshaking for the three lines involved and the data transfer (8-bit lines) will be shown shortly. First assume that while an ATN signal is still low (which means data lines may be used for commands and device addresses), the controller tells which device is the talker and which is the listener. Furthermore, assume the controller is also the talker and only one device is connected as a listener for this simplified case. One may think of the DAV line as being like a MASTER and both the NRFD and NDAC lines acting like a SLAVE; however, the NRFD line is used to tell the controller that DATA will be accepted from the bus (when it goes high), and the NDAC goes high as it accepts the data. See Figure 7.40.

This sequence of events is a full handshake and the system is now ready for the next operation.

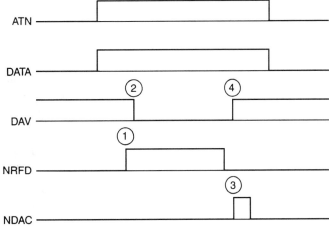

Figure 7.40 A timing diagram for an IEEE-488 bus with talker (controller/master) and listener

DIGITAL SYSTEMS

Many digital systems today involve the use of a microprocessor or a mini/micro computer someplace in the system. However, this subject area is much too lengthy to be included in a short review such as this. Usually one is either versatile in this area or has never been involved in the actual design of a problem solution using these devices. To attempt to study this subject for the first time for the sole purpose of answering questions on the PE examination is probably not the best use of one's time. For those who still would like a fast "run through" of this subject, but have little background in the area, an older text (still widely available), written at an introductory level, able to be read by the average EE in one evening, is one by Malvino. His chapters on SAP-1, SAP-2, and SAP-3 (Simple As Possible micro-computer) are perhaps the best and simplest introduction of the subject by any author.

Even though the understanding of microprocessors is fundamental in design-ing many digital systems, the subject of digital interfacing may be discussed without this detailed knowledge. In fact, digital problems on the PE examination may well address this hardware subject. The following few paragraphs will discuss encoders, coding, and A/D and D/A conversion.

Digital Encoders

Digital type encoders, whether used as error detectors in a control system or as a direct transducer for measurement, are available in a number of forms. Of course the simplest type for rotational measurement would be an incremental shaft encoder to measure displacement and velocity. Here the face-plate could be noth-ing more than a number of fine slots (cut radially in the face plate out from the shaft) that is capable of interrupting a light beam. Then, counting the number of light pulses from an optical detector would give the relative position of the shaft and the average velocity would be the number of counts per unit time. A digital counting circuit would easily give this information if the shaft only rotated in one direction. While this device works well, the disadvantage, in addition to the requirement to turn in one direction, is the possible loss of a reference position in case of power interruption or a "glitch" in a signal. A better encoder is one that is absolute in detecting a position. Consider a shaft face-plate that is labeled with the binary code as in Figure 7.41 (assume it is painted with reflective paint in concentric circles with each circle being detected with an optical reader).

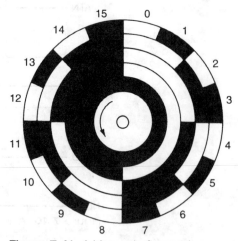

Figure 7.41 A binary shaft encoder

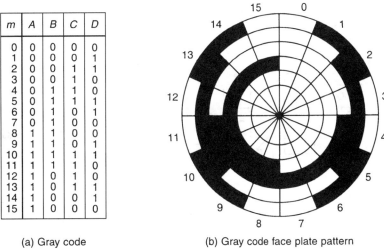

m	A	B	C	D
0	0	0	0	0
1	0	0	0	1
2	0	0	1	1
3	0	0	1	0
4	0	1	1	0
5	0	1	1	1
6	0	1	0	1
7	0	1	0	0
8	1	1	0	0
9	1	1	0	1
10	1	1	1	1
11	1	1	1	0
12	1	0	1	0
13	1	0	1	1
14	1	0	0	1
15	1	0	0	0

(a) Gray code   (b) Gray code face plate pattern

Figure 7.42 A 4-bit Gray code absolute shaft encoder

If each ring is painted in segments related to 2^n, it becomes a binary encoder. However, it should be noted that if the shaft's face-plate happens to have a number of segment changes being directly read by the detector(s), the digital code could become quite ambiguous at the border lines as a slight change in direction could make the output either a zero or a one for several of the bits. On the other hand, rather than using the binary code, another code, the Gray code could be used. This code has the advantage of not having more than one segment changing at a time as the segments pass under the optical detector (see Fig. 7.42).

Because the data from a Gray code encoder is not in either the natural binary or standard binary coded decimal format, it usually needs to pass through a Gray/binary code converter for processing. (Of course, if the data goes directly to a computer, it could be converted by software.) In practice, a 4-bit encoder divides the faceplate such that the absolute position may only be known by $360°/2^n = 360/16 = 22.5°$, which is far too great, therefore commercial encoders have at least 8-bits ($360/256 = 1.4°$) for a reasonable resolution. Consider how an absolute Gray encoder might be used in a control system in the following paragraph(s).

Gray to Binary Conversion

Assume a digital computer is available to control a partial analog system that drives a radar dish by its output shaft, and the position is detected by an 8-bit Gray encoder. As shown in Figure 7.43, the output of the Gray encoder passes through a Gray-to-binary converter to an I/O port of the computer. Assume the software has already been written for the call routines for the I/O 8-bit ports and other information critical to correct positioning of the dish. (This would include sign detection for clockwise or counterclockwise direction of the motor—that is, the plus or minus voltage output of the amplifier to the motor.) This system has the potential of generating several questions that might be on an examination. One such question could ask about getting the correct polarity for the servo amplifier from digital data, another question might concern the operation of the D/A converter itself, still another might be the operation of the I/O port.

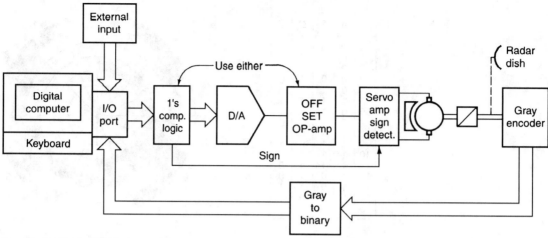

Figure 7.43 A digitally controlled system

By examining the Gray code (in Fig. 7.42a), it is easy to show that the converter logic circuit

1. passes the MSB (Most Significant Bit, or leftmost bit) directly through the converter unchanged,

2. for each other descending bit (on the Gray side), the natural binary is the same if the number of 1's to the left are even,

3. if the number of ones to the left is odd then the bit is changed.

The Gray-to-natural-binary converter and latch (could be a 74174 IC) may be designed as follows (see Figure 7.44): An 8-bit encoder, the Gray-to-binary converter is merely extended to convert the last 4-bits. The signal out of the I/O port, of course, is normally an 8-bit natural binary number and the D/A would convert it to an equivalent decimal number between 0 and 255; but, this is an increasing positive value and not suitable as an error voltage for the servo amplifier. There are several ways to overcome this deficiency. One such technique is to use the offsetting method (after the signal passes through the D/A converter) to realize

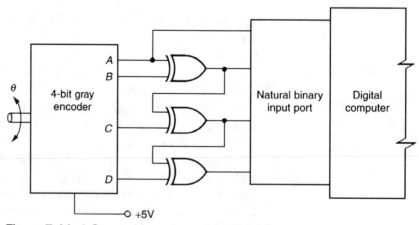

Figure 7.44 A Gray-to-binary converter with latch

that if a reference voltage is 30 volts, then byte 00000000 is equivalent 0 volts and byte 11111111 is equivalent to 30 volts, and the resolution is 30 volts/256 = 0.1172 volts per encoder division. However, by taking the output and offsetting the voltage with operational amplifiers (with their supply voltages being +/−15 volts), then the 255 data, being +30 volts, could be offset to +15 volts, the 127 data would be offset to 0 volts, and the 0 data would be offset to −15 volts out of the last op-amp.

Sign Convention

Another way of getting a +/− error voltage for a servo amplifier is to use the 1's complement method on all bits except the MSB; here the MSB would be considered as a sign bit with a 0 being positive and a 1 being negative. (The 1's complement method involves merely inverting each bit.) The logic circuit (see Fig. 7.45) is placed between the computer output port and the D/A converter and the sign bit is taken directly to the servo amplifier that is configured to change the sign of its output voltage by sensing the MSB. Of course the software would necessarily include an algorithm for operating on the data from the I/O ports and any other reference input for correct operation.

D/A Converters

For reviewing the digital systems area, both A/D and D/A converters should be included. However, because shaft encoding is one form of A/D conversion, this portion of the discussion will deal only with D/A converters. Most D/A converters convert directly from a natural binary code by switching successive weighted data bits to analog values by use of a summing junction. A typical op-amp circuit is used; although a number of other kinds of D/A converters exist, the successive weighted one is easy to understand and will be presented here. The operation of a "summer" for analog computers was discussed in the previous chapter and should be reviewed if necessary. Consider the circuit of Figure 7.46. Each bit is taken to be a weighted bit; for instance if SW(MSB) is ON, then the op-amp output is $1/2\ V_{ref}$, if only the SW(LSB) is ON for a 4-bit device the output is $1/16\ V_{ref}$. If V_{ref} is 16 volts, then the output may go (when all SW's are ON) to $(1/2 + 1/4 + 1/8 + 1/16)16 = 15$ volts. The functional diagram is shown in Figure 7.47 without the control lines.

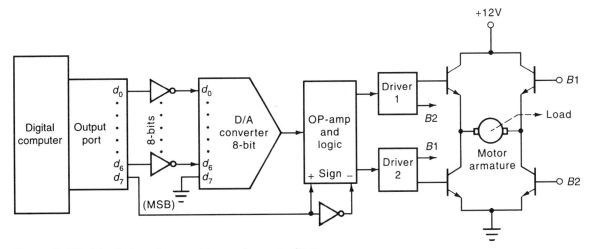

Figure 7.45 A technique for supplying a +/− error voltage

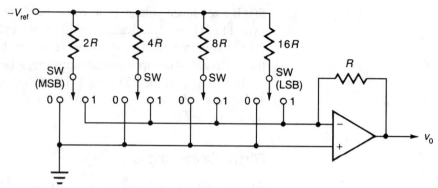

Figure 7.46 Analog portion of a basic D/A converter

Several of the specifications for a D/A need to be considered.

1. Accuracy—the difference between the expected value and the actual output value for a given digital input code, this should be less than +/– 1/2 the LSB. As an example, for an 8-bit converter, with a reference voltage of 16 volts, the LSB is 16/256 = 0.0625 volts, or the maximum error should be less than 1/2 that or 0.0312 volts.

2. Full scale—the FS output is the maximum output that may be obtained from the device; that is, if all of the inputs are 1's, then the output is $(2^n - 1)/2^n \times V_{ref} = 255/256(16) = 15.94$ volts.

3. Full Scale Range—the FSR is essentially the same as V_{ref}.

4. Resolution—the resolution is the reciprocal of the number of discrete steps (which is a function of the number of bits) and the total number of steps is $2^n - 1$. Expressed in percent is 0.392% (for the 8-bit converter).

5. Linear Errors—the deviation of the actual output from a straight line drawn through the idealized output for all steps. A special case, when all of the input lines are zero, is the actual output voltage (the "offset error voltage").

A D/A converter may be unipolar or bipolar as well as having a serial or parallel digital input. If the device is bipolar the output usually requires two op-amps, or, they may be built into the system. For bipolar operation, sometimes the MSB is the sign bit.

A very popular D/A converter, a DAC-08, has been and still is an industry standard. This DAC is an 8-bit fast (85 ns) current output (that is easily converted

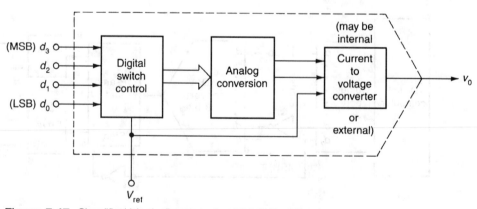

Figure 7.47 Simplified block diagram of a 4-bit D/A converter

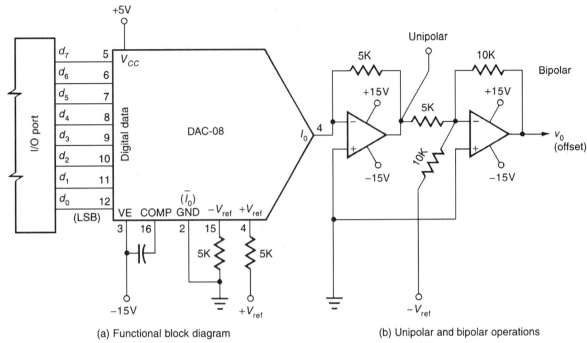

(a) Functional block diagram

(b) Unipolar and bipolar operations

Figure 7.48 A typical DAC-08

to a voltage with or without an external op-amp) that is unipolar; this unit may be converted to a bipolar type with an extra off-setting op-amp. For unipolar operation refer to Figure 7.48a, and as a bipolar device, refer to Figure 7.48b. Actually the output currents may be converted to output voltages directly without the use of any op-amps (for a high impedance load), by taking the voltage drop across resistors (500 ohms) to ground. The equivalent circuit for the DAC-08 is shown in Figure 7.49.

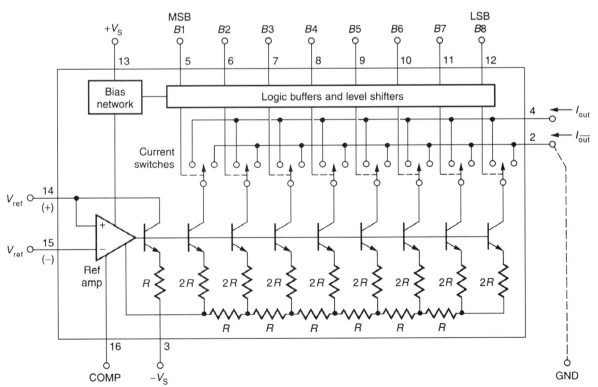

Figure 7.49 The DAC-08 equivalent circuit

Many problems for an examination can be generated requiring the use of A/D and/or D/A converters. One reference you may wish to bring to the exam is a manufacturer's data book on typical data conversion devices.

RECOMMENDED REFERENCES

Bartee, *Digital Computer Fundamentals*, 6th ed. McGraw-Hill, 1985 or later edition.

Byers, "The Complete PC Expansion-Bus Guide," *MicroComputer Journal*, formerly *ComputerCraft*, Vol. 1, No. 1, Jan./Feb., 1994.

Hill & Peterson, *Digital Systems: Hardware Organization and Design*, John Wiley & Sons, 3rd ed., 1987.

Malvino, *Digital Computer Electronics*, McGraw-Hill, 1977.

Mano, *Computer Engineering Hardware Design*, Prentice Hall, 1988.

Roth, *Fundamentals of Logic Design*, West Publishing Company, 1985 or later edition.

Communication Systems

This chapter includes transmission line theory, an introduction to antenna theory, and certain aspects of communications and signal processing theory. The reader is urged to have a formal text available for detailed proofs and theorems where needed. The review will begin with the representative cases of a transmission line. The study of transmission lines at lower frequencies (three-phase power transmission, etc.) was partially covered in Chapter 3.

The section on antenna theory is necessarily limited due to the high degree of specialty in this area. One would expect that the reader skilled in this field would have very little trouble with any examination problem. On the other hand, the reader unskilled in antenna theory would probably be well advised, except for the simplest of problems, to choose another problem area. This author recommends that only those readers with a good background in electromagnetic field theory use review time for this area of study.

The last section in this chapter deals with communication and signal processing at an introductory level. Again, there are many aspects of this subject that are too broad to cover in any detail. However, the introduction provided should jog one's memory in several aspects of this area. Several references will be given for further study.

TRANSMISSION LINES

For this portion of the review, most of the discussion will be based on the higher frequency transmission lines (open wire pair, coaxial cables, wave guides, etc.). In general, the common goal is to match the source impedance to the load through the transmission lines for maximum power transfer. For a closely matched source, transmission line, and load, any reflections on the line from the load will be limited. This discussion will begin with an idealized case of a transmission line being matched at both the sending and receiving end.

Basic Relationships

The input voltage at the sending end of the line is $v_1(t)$, while toward the receiving end it will be $v_2(t')$, where $t' = t - t_{delay}$. At the actual receiving end (or for notational purposes, the load end), $v_2 = v_r$. For a finite length, ℓ, the receiving voltage is $v_r = Kv_1(t')$; here the attenuation factor, K, will be less than unity because of the line losses. However, when working with sinusoidal quantities, rather than using the time delay expression, it is more convenient to express the attenuation factor as $e^{-\gamma\ell}$ with γ(Gamma) being a complex quantity (to be discussed later). It is desirable to have the transmission loss, TL, as near as possible to unity. However, TL increases exponentially with ℓ. This loss is given as $TL = 10e^{(\alpha\ell/10)}$. α is the attention constant and is given in either nepers per unit length or in dB per unit length (1 neper = 0.115 dB). TL may also be expressed in nepers or dB: $TL_{db} = 10 \log(P_{in}/P_{out}) = \alpha\ell$.

A typical T-section model for many transmission lines is shown in Figure 8.1.

A coax line may have one set of parameters while an open line may have another. For the open wire line (with wire spacing, b, and wire radius, a), the inductance per unit length, L', was previously developed (see the section "Line Inductance" in Chapter 1) and is given by

$$L' = (\mu_o/(\pi))\ln(b/a) = 4 \times 10^{-7}\ln(b/a) \text{ henrys/meter} \qquad \textbf{(8.1a)}$$

(where μ_o = permeability of free space = $4\pi \times 10^{-7}$).

The capacitance (see the section "Line Capacitance" in Chapter 1) is given by

$$C' = \pi\varepsilon/\ln(b/a) = 27.7/\ln(b/a) \text{ pf/m} \qquad \textbf{(8.1b)}$$

(where ε = dielectric constant of space = 8.85×10^{-12} MKS units).

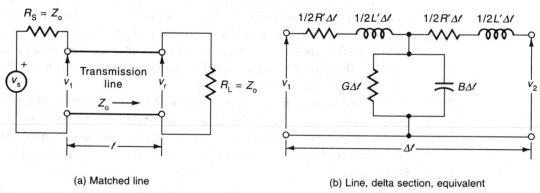

(a) Matched line   (b) Line, delta section, equivalent

Figure 8.1 A T-section model of a transmission line

The ac resistance, due to skin effect, may be needed, but if $\omega L >> R$, it is frequently neglected. For copper at 20°C, and for line spacing much greater than the wire diameter, an approximate equation for higher frequencies may be given as

$$R'_{ac} = 7.6a\sqrt{f} \times R_{dc} \text{ ohms/m} \quad \text{(See cautionary note.*)} \qquad (8.2)$$

For an open wire pair, the shunt loss, G, is essentially zero. The characteristic impedance can be greatly simplified to be

$$Z_o = 276 \log(b/a) \text{ ohms} \qquad (8.3a)$$

(For a coaxial cable, with air as a dielectric, the equivalent simplification is given by

$$Z_o = 138 \log(b/a) \qquad (8.3b)$$

where a = outside diameter of inner conductor and b = inside diameter of outer conductor.)

For a solid dielectric material, the conductance may be significant. Here the quality of the dielectric may be measured by means of the power factor. Because Y of the center leg of Figure 8.1b is given as $Y = G + j\omega C$, the power factor is given as

$$\text{PF} = G/\sqrt{G^2 + (\omega C)^2}$$

and, at higher frequencies if $\omega C > G$, then $G \to (\omega C) \times (\text{PF})$. As noted previously, if a transmission line is not lossless, the power loss in the line itself is not linearly proportional to the line length, but instead varies exponentially with length. Thus it is customary to express the line losses in terms of decibels per unit of length.

An example will be given for the power requirements of a transmission line. Assume both the source and load are properly matched and calculations for the various parameters have already been made. Suppose the power ratio calculations yielded $\alpha = 0.0025$ dB/m (recall, $\alpha_{dB}\ell = 10\log P_{in}/P_o$). Now assume that the available power source is only 5 mW, but 100 mW is needed at the receiving end, located 15 km away. Furthermore, assume two (properly matched) preamplifiers (power amplifiers) are available to boost the power, but each amplifier must have

---

*CAUTIONARY NOTE: The ratio of the actual resistance to that of zero frequency (dc) is not as simple as Equation 8.2 might suggest, and should not be used for lower frequencies. A graphical plot of a family of R_{ac}/R_{dc} curves is a function of both frequency and wire diameter, and all of the curves approach unity as the frequency falls toward zero.

The impedance of a conductor is, in general, complex (with real and imaginary parts) and the magnitude of these components is a function of frequency. The change of these quantities from their dc values is due to the skin effect; the current density is high near the surface and is minimal at the axis for higher frequencies. As an example, the skin effect for a round copper wire (with a 1 mm diameter) causes the current density at the axis to be less than half of the current density at the surface for $f = 100$ kHz. For a frequency of 1 Mhz the current density is almost zero at the axis, while the highest concentration is at the surface.

This skin effect penetration is usually designated as delta (δ). It may be shown that δ is related inversely to the square root of frequency. Thus, for higher frequencies, the ac wire resistance as shown in Anderson's text, *Electric Transmission Line Fundamnetals* (Reston Publishing Company, 1985, pp. 307, 312) may be given as $R_{ac} \approx a/(2\delta)R_{dc}$ where $\delta = 1/(\pi f\mu\sigma)^{1/2}$. Because of the complexities of the skin effect, this resistance usually can be found directly in tables or graphs in some standard EE handbooks.

Figure 8.2 Transmission line with preamplifiers

at least 2 μW for an input and deliver not more than one watt at the output to remain linear. Determine the power loss in the line and what the power gains of the amplifiers should be (see Fig. 8.2).

First determine power gains needed.

$$P_o/P_{in} = (PA_1 PA_2)/TL = (100 \text{ mW})/(5 \text{ mW}) = 20$$
$$\alpha\ell = (0.002 \text{ dB/m})(15 \times 10^3 \text{m}) = 0.0374 \times 10^3 = 37.4 \text{ dB}$$
$$TL = 10^{(37.4/10)} = 5{,}495 \approx 5{,}500$$
$$PA_{tot} = PA_1 PA_2 = TL(P_o/P_{in}) = 5.5 \times 10^3 (100\text{m W})/(5\text{mW}) = 110 \times 10^3$$
$$(PA_{dB} = 10 \log PA_{tot} = 10 \log 5.5 \times 10^3 = 37.4 \text{ dB})$$

Note that a preamplifier is needed to boost the source; choose $PA_1 = 200$ (which keeps its output well within the linear range, $200 \times 5\text{m W} = 1 \text{ W}$).

$$P'_o = (P_{in} \times 200)/TL = (5 \times 10^{-3} \times 200)/(5.5 \times 10^3) = 0.182 \text{ mW}$$

This new value of power output is much less than the 100 mW required, but well above the noise level of the second preamplifier,

$$PA_2 = (100 \times 10^{-3})/(0.182 \times 10^{-3}) = P_{o(\text{req'd})}/P_{o(\text{act't})} = 550 \text{ (Gain)}.$$

The following is just a check to see if the final total power output is 100 mW:

$$P_o = P_{in}PA_1 PA_2/TL = (5 \times 10^{-3} \times 200 \times 500)/(5.5 \times 10^3) = 100 \text{ mW}$$

Lossless Transmission Lines

Consider a transmission line that is completely lossless (a zero dissipation line). Then Z reduces to $j\omega L$ and Y reduces to $j\omega C$, and the line characteristic impedance becomes

$$Z_o = \sqrt{Z/Y} = \sqrt{L/C} \text{ ohms (purely resistive)}.$$

The propagation constant, γ, may be shown to be

$$\gamma = \sqrt{(j\omega L)(j\omega C)} = j\omega\sqrt{LC} = \alpha + j\beta \rightarrow j \quad \text{as} \quad \alpha = 0$$
$$\text{and} \quad \beta = \omega\sqrt{LC}.$$

If the line is terminated into its characteristic impedance, Z_o (equals R_o), then the line is properly matched and maximum power is delivered to the load. For this condition, the voltage and the current (given in rms values) may be found at any point, d, back from the receiving end from the following two equations:

$$V = V_R\cos(2\pi d/\lambda) + jI_R R_o\sin(2\pi d/\lambda) \tag{8.4a}$$
$$I = I_R\cos(2\pi d/\lambda) + j(V_R/R_o)\sin(2\pi d/\lambda) \tag{8.4b}$$

where V_R and I_R are measured at the receiving end, and λ is the wavelength, where $\lambda = (3 \times 10^8)/f$ in free space.

On the other hand, if the load is mismatched at the receiving end, a reflected wave is sent back along the line (the original wave from the source end is called the incident wave). The reflection coefficient, ρ (or sometimes given as Γ), is the ratio of the reflected and the incident voltage waves as measured at the load, and is given by

$$\rho = (Z_R - Z_o)/(Z_R + Z_o). \qquad (8.5)$$

If all of the energy is rejected at the load, the reflected wave is as large as the incident wave; however, because the load is entirely reactive and cannot consume energy, it can store energy for part of a cycle and return it to the line, but not in phase with the incident wave. As an example, if Z_r is entirely capacitive reactive, I_r leads V_r by 90 degrees and the reflection coefficient, ρ, will have an angle of –90 degrees. The measured voltage at any point along the line will be the vector sum of the incident and reflected waves, called a "standing wave." Also, if the load impedance is not matched to the line, a resulting standing wave will be present. These standing waves have maximum and minimum values for both currents and voltages. This standing wave has a ratio, SWR (or frequently abbreviated as S), and is defined as

$$\text{SWR} = S = |V_{max}/V_{min}| = |I_{max}/I_{min}| = (1 + |\rho|)/(1 - |\rho|). \qquad (8.6)$$

For an unmatched load, the impedance of the line varies, depending on the distance, d, back from the load end. It is easily calculated at any point by dividing the voltage by the current at that point (refer to Eq. 8.4a,b). For a resistive load (here $R_r = R_L$) if R_L is greater than R_o, the calculated resistance at that point falls from R_r to a minimum at the quarter wave length point. At the same time, the impedance becomes complex with a capacitive reactance component; in other words, at a point d, the load begins to look like a series impedance made up of R and $-jXc$! As soon as d passes the quarter wave length point, the capacitive reactance component suddenly changes to inductive reactance up until a half wave length is reached, and continues switching back and forth every quarter wave length.

Open or Short Circuit Lines

Assume, instead of terminating a transmission line in its matching resistance, the leads are open circuited or, on the other hand, are shorted together. At the termination point, for the open circuit case, the current is zero, while for the shorted case the voltage is zero. The effect is, of course, to set up standing waves. For the shorted case, a voltage standing wave will appear on the line with its zero value at the termination end and its maximum value located a quarter wave length back along the line. For current, the maximum will appear at the termination end, and the minimum, a quarter wave back. The effect is that as one moves back away from the shorted end, the impedance at d appears as a pure inductive reactance; for a d greater than a quarter wave length, the reactance switches to capacitive. This "shorted stub" effect may be taken advantage of by helping to match an otherwise unmatched load.

Stub Matching Procedure

To describe the following stub matching procedure (and also for those wishing to use the SMITH chart), it will be helpful to use per unit or normalized values for

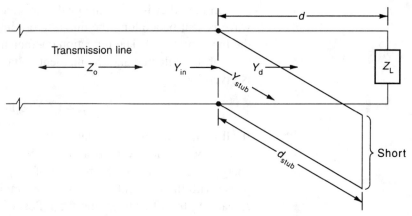

Figure 8.3   Placement of a shorted stub

impedances. These values are nothing more than all the individual impedances divided by the characteristic impedance, Z_o; these normalized values are usually indicated by lowercase letters. For instance, since $z_o = Z/Z_o = 1.0$ then $r_r = R_r/Z_o$.

Suppose one wants to match a two-wire transmission line to a load impedance of another value (its source and line impedance are equal, $Z_s = Z_o$). If one measured back a distance, d, from a load (see Fig. 8.3) and paralleled this with a shorted transmission line of the same Z_o, and length d_{stub}, hopefully this parallel combination would have a combination impedance of Z_o. The parallel combination at this point, d, is most easily analyzed by using admittances. The total input admittance is given by

$$Y_{in} = Y_d + Y_{stub} = Y_o = 1/Z_o, \qquad (8.7)$$

or, using normalized values (*i.e.*, $y_b = Y_b/Y_o$, etc.), the equation reduces to

$$1 = y_d + y_{stub}.$$

Here the normalized admittance of the stub is purely imaginary; thus y_d must equal

$$y_d = 1 + jb_{stub} = 1 + jb_d; \quad \text{and} \quad y_{stub} = -jb_d.$$

From the above equations it may be shown that the normalized impedance looking toward the load end at d is given by

$$z_d = [(r_L + jx_L) + jk]/[1 + j(r_L + jx_L)k], \qquad (8.8)$$

where $k = \tan \beta d$. The normalized admittance looking from d to the load is

$$y_d = 1/z_d = g_d + jb_d \qquad (8.9a)$$

and is given by

$$g_d = [r_L(1 - x_L k) + r_L k(x_L + k)]/\left[r_L^2 + (x_L + k)^2 \right] \qquad (8.9b)$$

$$b_d = \left[r_L^2 k - (1 - x_L k)(x_L + k) \right]/\left[r_L^2 + (x_L + k)^2 \right]. \qquad (8.9c)$$

Because g_b needs to be unity for perfect impedance matching, these equations can be solved directly for k (depending on whether r_L = unity).

$$k = [1/(r_L - 1)][x_L \pm \sqrt{r_L \left[(1 - r_L)^2 + x_L^2 \right]}, \quad \text{for} \quad r_L \neq 1 \qquad \textbf{(8.10a)}$$

$$k = -x_L/2, \quad \text{for} \quad r_L = 1 \qquad \textbf{(8.10b)}$$

The distance back from the load, d, to the parallel connection is given by

$$d = [1/(2\pi)][(\tan^{-1} k)]\lambda, \quad \text{for} \quad k \geq 0 \qquad \textbf{(8.11a)}$$

$$d = [1/(2\pi)](\pi + \tan^{-1} k)]\lambda, \quad \text{for} \quad k < 0 \qquad \textbf{(8.11b)}$$

and the stub length is then

$$d_{\text{stub}} = [1/(2\pi)][\tan^{-1}(1/b_b)]\lambda, \quad \text{for} \quad b_d \geq 0 \qquad \textbf{(8.12a)}$$

$$d_{\text{stub}} = [1/(2\pi)][\pi + \tan^{-1}(1/b_d)]\lambda, \quad \text{for} \quad b_d < 0. \qquad \textbf{(8.12b)}$$

From these equations, the placement and length of stub may be determined.

For instance, a 50 ohm transmission line is to be terminated in a 75 ohm load, and the signal source is operating at 100 MHz. The reflection coefficient is found to be $\rho = (Z_L - Z_o)/(Z_L + Z_o) = (75 - 50)/(75 + 50) = 0.2$. Or, the VSWR is given by $(1 + |0.2|)/(1 - |0.2|) = 1.5$. (These calculated values of ρ and VSWR are redundant for the particular formulas used for the solution of this problem, but are nevertheless, interesting data.) Here, the goal is to select a distance d away from the load to attach a shorted line with a length d_{stub}. The solution involves first finding the normalized values of the load and then finding k,

$$z_L = r_L = R_L/Z_o = 75/50 = 1.5.$$
$$k = (\text{see Eq. 9.10a}) = [1/(1.5)] \sqrt{0 + 1.5(1 - 1.5)^2 + 0} = 1.225$$

Then calculate the d's:

$$d = (\text{see Eq. 9.10a}) = [1/2\pi)][\arctan(1.225)](3.0) = [0.159][50.8°](3)$$
$$= [0.159][0.886 \text{ rad}](3) = 0.424 \text{ m}$$

To calculate the stub length, b_d is needed, which, from Equation 8.9c,

$$b_d = [(1.5)^2(1.225) - (1 - 0)(0 + 1.225)]/[(1.5)^2 + (1.225)^2] = 0.409.$$

Then d_{stub} from Equation 8.12a is

$$d_{\text{stub}} = [1/(2\pi)][\arc \tan(1/0.409)](3) = 0.567 \text{ m}.$$

For the interested reader, another way of solving the preceding problem (and making use of the redundant data), is a method of working from the first voltage minimum away from the load end, d_{min}, and calling this point reference. This (VSWR) voltage minimum will be exactly a quarter wave back from the load *if the load is a pure resistance.* Then the formulas needed for placing the stub toward the load but away from the closest voltage minimum (*i.e.*, $d_{min} - d = d_x$) are

$$d_x = [\cos^{-1}(\text{SWR} - 1)/(\text{SWR} - 2)](\lambda/4\pi) \text{ meters} \tag{8.13a}$$

$$d_{stub} = \frac{\lambda}{(2\pi)} \tan^{-1}\left[\frac{\sqrt{\text{SWR}}}{(\text{SWR}-1)}\right] \text{m.} \tag{8.13b}$$

For the previous problem with $Z_o = 50\Omega$ and $Z_L = 75\ \Omega$, the distances are

$$d_x = \left[\cos^{-1}\frac{(1.5-1)}{(1.5+1)}\right]\left(\frac{3}{4}\pi\right) = [78.5° \times \pi/180°](0.239) = 0.327 \text{ m}$$

$$d_{stub} = \frac{3}{2}\pi \tan^{-1}\left[\frac{\sqrt{1.5}}{(1.5-1)}\right] = 0.565 \text{ m.}$$

A number of restrictions and idealized conditions were indicated throughout this short review. A somewhat more practical approach to these kinds of problems makes use of the SMITH* Chart and the reader is encouraged to study the following properties.

Smith Chart Properties

A Smith chart is used in many instances to determine the parameters of a lossless transmission line. The Smith chart has the following properties:

- All possible values of impedance are contained inside the outer circle of unit radius.

- βs increments are indicated around the outer edge of the chart in terms of wavelengths.

- A straight edge pivoted at the center and marked in terms of S serves as a distance coordinate to any point on the chart and has the effect of adding constant-S circles to the chart without actually complicating the figure with additional lines.

- The impedance of a transmission line may be read at any point on the appropriate S-circle.

- The point at the center of the chart represents the impedance of the line terminated in its characteristic impedance, where $Z/R_0 = 1$ for all distances.

- The point at the extreme left of the resistance r_a axis represents a short circuit (zero impedance), and the point at the extreme right represents an open circuit (infinite impedance).

- The outer circle represents $S = \infty$.

*"Smith" is a registered trademark of Analog Instruments Co., Box 808, New Providence, NJ 07974. The Smith Chart is reproduced by courtesy of Analog Instruments.

- The chart may be used for admittance as well as impedance, the r_a and x_a axes becoming g_a and b_a axes, with the convention that capacitive susceptance is positive (above) and inductive susceptance is negative (below); the leftmost point is then an open circuit (zero conductance), and the rightmost point is a short circuit (infinite conductance).

- V_{min} occurs on the real axis. When using impedances, V_{min} occurs on the left half; when using conductances, V_{min} occurs on the right half.

Before discussing these properties, refer to the unmarked Smith chart in Figure 8.4.

NAME	TITLE	DWG. NO.
SMITH® CHART FORM ZY-01-N	ANALOG INSTRUMENTS COMPANY, NEW PROVIDENCE, N.J. 07974	DATE

NORMALIZED IMPEDANCE AND ADMITTANCE COORDINATES

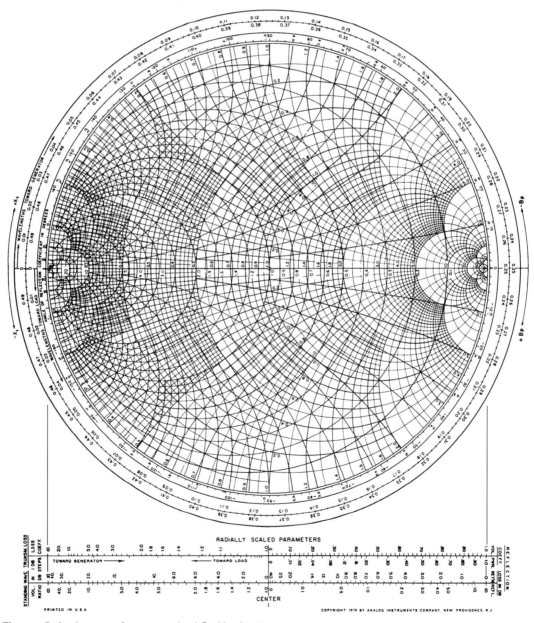

Figure 8.4 A copy of an unmarked Smith chart

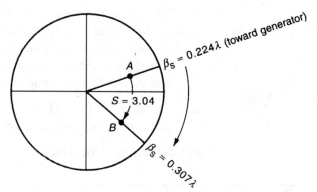

Figure 8.5   Construction lines on SMITH chart

As an example, assume one is interested in finding the normalized sending impedance, Z_s, using a Smith chart, given that,

$$Z_R/R_o = 2.5 + j1.1 \ \Omega; \text{ line length} = 0.38.$$

(*Note:* Smith chart problem solutions require a compass and straight edge.) The solution may be found as follows (refer to Fig. 8.5):

1. Locate load point A at $2.5 + j1.1$ on Smith chart.

2. Draw constant-βs line from origin through A to outer circle reading 0.224λ (toward generator).

3. Calculate line length in terms of λ

$$\frac{30°}{360°} = 0.083\lambda$$

and move this distance toward generator along outer circle to the point $(0.224 + 0.083)\lambda = 0.307\lambda$.

4. Draw another constant-βs line from origin to 0.307λ on outer circle.

5. Draw a portion of a constant-S circle (center at origin) from point A to point B, where it intersects with the constant-βs line drawn in step 4.

6. Read normalized input impedance at point B as $Z_s/R_0 = 1.59 - j1.37$ (capacitive reactance). This is the answer.

For more details on Smith chart usage, please refer to the example problems at the end of the chapter.

ANTENNAS

Problems dealing with antennas cover a very broad range of subjects usually dealt with under the heading of electromagnetic field theory. However, the subject is included in this chapter following transmission lines as it seems to "fit" better in this kind of limited review. For those well versed in electromagnetic field theory, probably a good short reference chapter is the last chapter in the previously referenced book by Anderson; for those not so well prepared, the *AARL Antenna Book** is widely available. Of course any of the full text books, such as Kraus's *Electromagnetics* or Ramo's *Fields and Waves in Communication Electronics*, present the basic fundamentals in detail. In any case, it is recommended that one study both electromagnetic field theory and transmission lines before starting this review. Most problems involving antennas assume a basic knowledge of at least antenna impedance, patterns and kinds of antennas, power, and of course, frequency range.

Before reviewing some of these concepts, one should recall the following useful information:**

1. Impedance of free space = 377Ω.

...ic means identical in all directions.

...on resistance of a half-wave ($\lambda/2$) dipole = 73.26Ω and is best fed ...72 Ω cable.

...f a $\lambda/2$ dipole = unity (reference) when the dipole is oriented to produce ...ximum gain in the same direction as the actual antenna.

...e area of a sphere = $4\pi r^2$. Surface area of a cylinder = $2\pi rb$.

...of a microwave antenna is given by the formula

$$G = \frac{4\pi A}{\lambda^2}$$

...avelength $= \dfrac{v}{f} = \dfrac{3\times10^8}{f}$ meters

...fective area $= \eta A_{\text{actual}}$

...ntenna efficiency

...er gain of an actual antenna is the ratio of the Poynting vector produced by ...actual antenna in a particular direction to the value of the Poynting vector ...erated in all directions by an isotropic source of power.

...m the Poynting vector relationship (see Eq. 8.15), the output power density ...y be expressed for an isotropic source or for far field (where W = rate of ...rgy flow per unit area) as

$P_{r\text{-iso}} = $ [Poynting vector relationship] $= \text{W}/(4\pi r^2) = P_{\text{incident}}$ watts/m^2.

...ntenna Directive Gain (in dB) $= 10\text{Log}(P_{r\text{-actual}}/P_{r\text{-iso}})$

$$= 10\,\text{Log}[(4\pi r^2)/W_{r\text{-iso}}]P_{r\text{-actual}}$$

...r example, a half-wave dipole antenna (for the maximum direction, $\theta = 90°$), ...$= 1.64 = 2.15$ dB. (Reference p. 649 of Ramo, Whinnery, and VanDuzer's text.)

...books, published by the American Radio Relay League, should be available at most stores ...l amateur radio equipment or may be ordered directly from ARRL Headquarters, 225 Main ...Newington, CT 06111; phone number 860-594-0200.

...e of this information is reprinted with permission from *Electrical Engineering: Referenced ...v for the PE Exam*, Kaplan AEC Education.

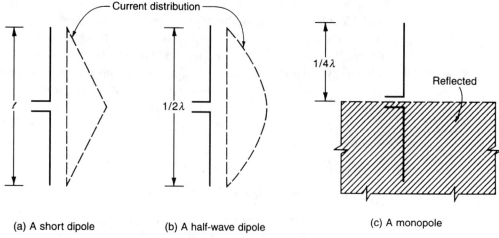

(a) A short dipole   (b) A half-wave dipole   (c) A monopole

Figure 8.6 Center-fed (thin wire) dipole antennas

Dipole Configuration(s)

Basic dipole antennas are usually referred to as full-wave, half-wave, and quarter-wave (or shorter types), or sometimes as folded-wave type. In addition, the method of how the antenna is connected to its source (such as center-fed) needs to be considered. For a center-fed short (much less than a wave length), the current distribution may be considered to be a maximum at the center and tapering off (linearly) to zero at the ends (see Fig. 8.6a). On the other hand, for a half-wave dipole (quarter-wave each side of the center-fed point), the current distribution is sinusoidal (see Fig. 8.6b). A monopole antenna (see Fig. 8.6c) may be thought of as a center-fed dipole with half of the dipole above ground and the ground itself as an ideal reflecting image of the upper half (but unfortunately radiating only half the power and half the radiation resistance of the short dipole type).

Impedance and Power Relationships

The impedance of the antenna is the RF voltage current ratio that varies along the antenna, and the ratio is the impedance at that particular point. The impedance also depends on the conductor size in relationship to the frequency; the capacitance per unit length increases and inductance per unit length decreases for increasing conductor diameter.

As indicated in item 1, the impedance of free space, or intrinsic impedance, is given as 377Ω, which is $\sqrt{\mu_0/\varepsilon_0}$. It also may be shown that, for a differential length antenna,

$$\sqrt{\mu_0/\varepsilon_0} = \frac{|E|}{|H|} = 120\pi = 377\Omega. \tag{8.14}$$

Because E and H are normal to each other and in time phase, for a small area of a sphere, the radiating wave appears as a plane wave for large values of a radius (with the source at the center of the sphere). The radiated power may also be calculated (for the larger radius, known as "far" field) from the flow of energy by considering Poynting's vector, thus the power density in an electromagnetic field is

$$S = P = E \times H = P_r a_r + P_\theta a_\theta \text{ (watts/meter}^2\text{)}. \tag{8.15}$$

The radiated power for a short dipole antenna becomes (for a "point source", $P_\theta a_\theta \approx 0$),

$$P = \int_s S_{avg} \cdot ds \text{ watts.} \qquad (8.16)$$

(Here S_{avg} is the average Poynting vector in watts/meter^2.)

And, for a short dipole antenna (assuming no losses), the power radiated must be equal to the average power delivered to the antenna terminals. This average power is $(1/2)\, I_o^2 R$ where I_o is the peak amplitude of the current. Thus the radiation resistance may now be defined as

$$R = 2P/I_o^2 \, \Omega. \qquad (8.17a)$$

With further manipulation of Equation 8.17a, one may show that for a short dipole for far field, the radiation resistance is

$$R = 80\pi^2 (\ell/\lambda)^2. \qquad (8.17b)$$

(Here ℓ is the length of the dipole in meters and λ is the wavelength in meters.) As a short example, the radiation resistance of a dipole of 1/8 wavelength may be found directly from Equation 8.17b as

$$R = 80\pi^2 (1/8)^2 = 12.3 \ \Omega.$$

For maximum power transfer between the feeding line and the antenna, the impedance is crucial; for a mismatch, the reflected power increases as the mismatch increases. At a particular operating frequency the antenna may be at resonance and the radiated power may be optimal and, of course, the feeder line impedance should match that of the antenna.

Power Measurements

Power measurements can often give enough information to allow one to calculate antenna radiation power and/or field strength. Consider, as an example, the power measurement for a vertical antenna at an AM broadcast transmitting station. Assume the antenna has an effective half-power radiation cross section configuration as shown in Figure 8.7. Further assume that at 75 meters from the antenna, a field strength meter would read the maximum rms value of field strength (currently unknown).

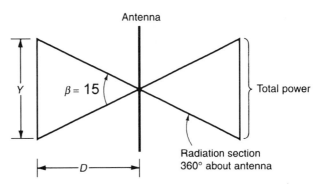

Figure 8.7 Effective half-power radiation cross section

The total value of the time-average power that the antenna radiates is known to be 7.71 kilowatts. Because the total power is given, one may then calculate the field strength as follows:

At 75 meters away from the antenna, the field strength is designated as $E_{\text{rms-max}}$, and the half-power point voltage/meter will then be $E_{\text{rms}} = E_{\text{rms-max}}//2$ and the power density may be expressed as

$$P_{\text{Density}} = |E|^2/Z_{\text{Free Space}} = (|E_{\text{rms}}|)^2/377.$$

From the geometry of the configuration, the area of the cylinder of revolution is determined as

$$\tan \beta/2 = 0.5(Y/75) = \tan 7.5°, \; Y = 19.74 \text{ m},$$
$$\text{Area}_{\text{cylinder}} = 2\pi DY = 2\pi 75 \times 19.74 = 9{,}302 \text{ m}^2.$$

Knowing the total power, the field strength may be determined by

$$P_{\text{Total}} = 7710 \text{ watts} = \text{Area}_{\text{cylinder}} P_{\text{Density}} = 9{,}302 \, (|E_{\text{rms}}|)^2/377, \; (|E_{\text{rms}}|)^2 = 312.5,$$
$$|E_{\text{rms}}| = /312.5 = 17.7 \text{ V/m}. \quad |E_{\text{rms-max}}| = /2E_{\text{rms}} = /2 \times 17.7 = 25 \text{ V/m}.$$

On an examination, after working out an engineering solution, it is always helpful to know what an approximate answer should be; many "simplified" formulas are presented in the *ARRL Antenna Book*. For instance, for a half-wave dipole antenna (in free space), the length is approximately

Length (in feet) = 495K/f(MHz),

where K is near enough to unity for frequencies below 30 MHz and varies downward depending on other factors for higher frequencies.

Of course many other aspects of antennas and antenna patterns could and probably should be studied for PE examination. For these, the reader is referred to the practice problems and a more complete text on the subject.

COMMUNICATIONS AND SIGNAL PROCESSING

In any study of communications systems, the subject of signal processing must be considered. This subject area involves spectral analysis, line noise and distortion, modulation, and sampling. A functional block diagram description is shown in Figure 8.8. These functions include both analog (usually sinusoidal) and digital (usually pulsed signals) ways of transmitting information.

As an example of the concept depicted in Figure 8.8, assume the input is an 8-bit digital data signal and the desired output is the same. If the transmission line happens to be a single telephone line (with some kind of clocking signal), the input processor could be a D/A converter and the output then would be an A/D converter

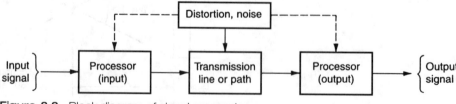

Figure 8.8   Block diagram of signal processing

Figure 8.9 An oversimplified data link (poor solution)

(see Fig. 8.9). While this technique would work, it is intuitively a poor solution. On the other hand, if the input processor was a parallel-to-serial converter and the output processor was a serial-to-parallel converter (both with proper synchronous or asynchronous coordination), the solution is intuitively better.

Here, these kinds of converters could be shift-registers and may be referred to as UARTs (universal asynchronous receiver/transmitter). Of course, for this simplified example, one is trading eight transmission lines for one, with the penalty of requiring at least eight (or more, depending on the protocol) time periods for one, for transmission of one byte of data. At any rate, the signals, along with interference and noise, need to be analyzed for a good engineering solution.

Fourier Series Analysis

Periodic Signals

If the signals are analog in nature and are periodic, then, of course, one may convert them to sinusoidal components by using the Fourier series method for further analysis. This spectral analysis technique involves the summing of the sinusoids. Recall, a function is periodic if $f(t) = f(t + T)$ and the Fourier series is then given by

$$\begin{aligned} f(t) &= a_o + a_1\cos\omega_o t + b_1\sin\omega_o t + a_2\cos 2\omega_o t + b_2\sin 2\omega_o t + \cdots \\ &\quad + a_n\cos n\omega_o t + \cdots b_n\sin n\omega_o t, \\ &= a_o + \Sigma(a_n\cos n\omega_o t + b_n\sin n\omega_o t). \end{aligned} \tag{8.18}$$

For these equations the coefficients are given by

$$a_o = (1/T)\int f(t)\,dt, \tag{8.19a}$$

$$a_o = (1/T)\int f(t)\cos n\omega_o t\,dt, \tag{8.19b}$$

$$b_n = (2/T)\int f(t)\sin n\omega_o t\,dt. \tag{8.19c}$$

Of course the average value a_o is zero if x-axis symmetry is present, a_n is zero for y-axis symmetry [with $f(-t) = -f(t)$], and b_n is zero for even symmetry [with $f(-t) = f(t)$]. For determining the line spectrum for a periodic signal, one only needs to find the complex values of the coefficients, c_n, along with the resulting phase as in the following equation (Eq. 8.20).

$$c_n = \sqrt{a_n^2 + b_n^2},\ \Phi_n = -\tan(b/a). \quad \text{(See note.)} \tag{8.20}$$

Note: If c_n is derived from its exponential equation form (from $t = -\infty$ to $t = +\infty$), then Equation 8.20 is divided by 2; however when plotting the line spectrum, the c_n term (as found from the exponential form) is multiplied by 2 so the results are the same.

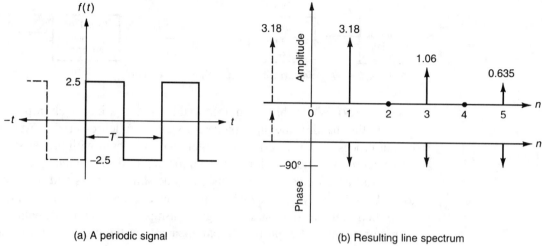

(a) A periodic signal   (b) Resulting line spectrum

Figure 8.10 Spectral analysis for a periodic signal

As an example, consider the periodic signal in Figure 8.10a and the resulting line spectrum of Figure 8.10b. Here, for the given square wave, the average value $a_o = 0$ and $a_n = 0$, b_n will equal c_n and $f(t)$ may then be found as

$$f(t) = (10/\pi)[\sin 1\,\omega_o t + (1/3)\sin 3\omega_o t + (1/5)\sin 5\omega_o t + \cdots (1/n)\sin(n\omega_o t)],$$

or may be written (because its line spectrum is normally plotted as a cosine function) as

$$f(t) = (10/\pi)[\cos(\omega_o t - 90°) + (1/3)\cos(3\omega_o t - 90°) + (1/5)\cos(5\omega_o t - 90°) + \cdots].$$

As a short example, assume one is remotely measuring a very small sinusoidal signal from a sensing device and wishes to record the signal on a recorder some distance away. Further, assume a preamplifier is connected to the measuring device but introduces harmonics to the original signal before it passes through a line to the recorder. As the signal passes through the long line, assume some noise gets on the line before it reaches the recorder. Figure 8.11 shows the spectral lines one encounters in the measurement system. The low-pass filter should have its cutoff frequency

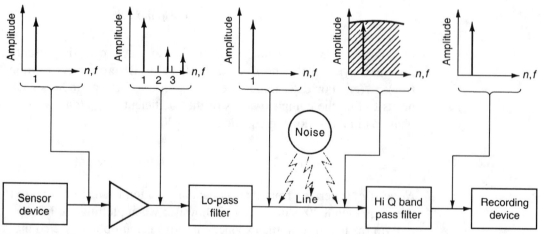

Figure 8.11 Recording a signal from a remote location

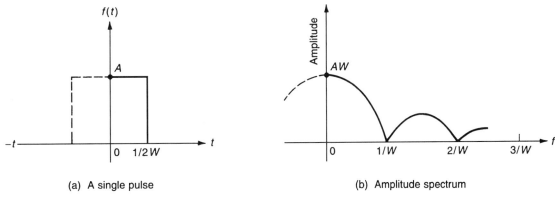

(a) A single pulse

(b) Amplitude spectrum

Figure 8.12 A single pulse with its amplitude spectrum

somewhat higher than the measurement frequency and, of course, the high-Q band-pass filter may be a turned one that attenuates spectral lines of both sides of the signal frequency. Most systems where the signals are periodic are straightforward to analyze, while nonperiodic ones may be quite difficult to calculate.

Nonperiodic Signals

In general, a series of digital nonperiodic signals does not lead to a simple Fourier analysis; however, a few generalities may be drawn. First consider a single pulse as shown (see Fig. 8.12) with a width, W, and an amplitude, A. One could consider this pulse as part of a periodic pulse train with a time period T approaching infinity (which makes all of the pulses vanish except the one centered at the origin). Here the amplitude lines begin to merge and, rather than lines, one could draw continuous lines or an envelope to represent a continuous amplitude spectrum. This means that the signal energy is spread over a range of frequencies rather than individual spectral lines. One author* asserts that the amplitude of the frequency spectrum for a pulse centered at $f = 0$ is approximately equal to the net area of the pulse itself. Thus the amplitude on the frequency scale is the amplitude on the time scale multiplied by the pulse width; these spectra envelopes become smaller as frequency increases and are known as low-pass signals. These kinds of nonperiodic signals are better suited to analysis by the Fourier transform and are covered in more detail in books on communication theory; however, for an introduction, an especially good reference is one by Nilssen (as previously noted elsewhere).

Modulation

Modulation normally involves two signals, the carrier frequency (usually in the RF spectrum), f_c, and the information signal (usually audio signals), f_s. For amplitude modulation, the magnitude of the carrier is related to the instantaneous value of the signal frequency such that

$$v_{\text{am}}(t) = (V_{c\text{-max}} + kV_{s\text{-max}}\cos 2\pi f_s t)\cos 2\pi f_c t$$
$$= V_{c\text{-max}}(1 + m\cos 2\pi f_s t)\cos 2\pi f_c t. \qquad \textbf{(8.21)}$$

---

*Carlson, *Electrical Engineering Concepts and Applications*, 2nd ed., Addison-Wesley Publishing Company, 1990, p. 475.

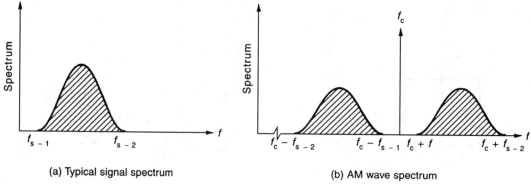

(a) Typical signal spectrum   (b) AM wave spectrum

Figure 8.13   Typical spectrum for AM signals

Here, the factor $m (m = V_{s\text{-max}}/V_{c\text{-max}})$ is called the modulation index and should be kept to less than unity to avoid distortion. The AM signal may be expanded and expressed as

$$v_{AM}(t) = V_{c\text{-max}}\cos 2\pi f_c t + (1/2)mV_{c\text{-max}}[\cos 2\pi (f_c + f_s)t$$
$$+ \cos 2\pi (f_c - f_s)t]. \qquad (8.22)$$

Thus the AM signal has two varying sidebands and one carrier signal. Since the varying signal may have a frequency spectrum as shown in Figure 8.13a, the overall frequency spectrum will be that of Figure 8.13b.

On the other hand, another category of modulation is called the product kind and, of course, is the product of the carrier and the information signal. Here, assume the peak amplitude of the carrier is set at unity and $f_s \gg f_c$, then

$$v(t) = (V_{c\text{-max}}\cos 2\pi f_s t)(\cos 2\pi f_c t)$$
$$= (V_{s\text{-max}}/2)[\cos 2\pi (f_c - f_s)t + \cos 2\pi (f_c + f_s)t]. \qquad (8.23)$$

This resulting signal contains only the sidebands and is referred to as a suppressed carrier signal. Assume this information signal is a relatively low fixed frequency, then a plot of the spectral frequency is shown in Figure 8.14.

Now assume both the amplitude and frequency of the information signal vary, then the spectral relationship may be depicted as in Figure 8.15.

As a simple example, suppose a 10 kHz (f_{c1}) carrier is product modulated by a 500 Hz (f_{c1}) signal. The resulting signal passes through a high-pass filter that has a sharp cutoff frequency of f_{c1}. The output of this filter is a new signal(s) that, in turn, will product modulate a second carrier of 100 kHz (f_{c2}). The final output

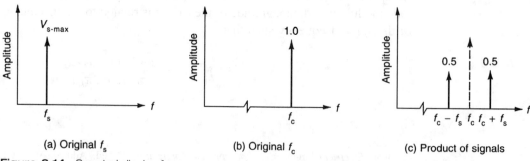

(a) Original f_s (b) Original f_c (c) Product of signals

Figure 8.14   Spectral display for a suppressed carrier system

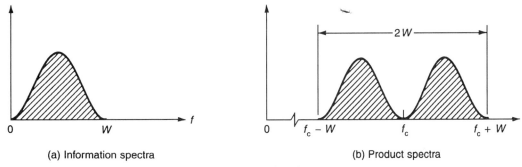

(a) Information spectra   (b) Product spectra

Figure 8.15   Spectral display varying information signal

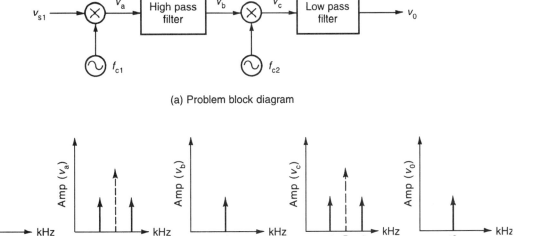

(a) Problem block diagram

(b) Frequency spectrum

Figure 8.16   Example problem solution

is now passed through a low-pass filter that has a sharp cutoff of f_{c2}. Figure 8.16a shows the block diagram for this example. What is/are output frequency(ies) of the last filter? Of course, for this simple example, one is merely involved with bookkeeping in obtaining the solution (Figure 8.16b).

For these kinds of problems, it is assumed the reader is already comfortable with various filter designs and has some background in the various modulation techniques including frequency modulation (FM).

RECOMMENDED REFERENCES

Anderson, *Electric Transmission Line Fundamentals,* Reston Publishing Company, Inc., 1985, Chapter 14.

Bentley, *Electrical Engineering: Referenced Review for the PE Exam*, Kaplan AEC Education, 2004.

Carlson, *Electrical Engineering Concepts and Applications,* 2nd Ed., Addison-Wesley Publishing Company, 1990, p. 475.

Kraus and Fleisch, *Electromagnetics,* 5th Ed., Mc-Graw Hill, 1999.

Ramo et al., *Fields and Waves in Communication Electronics,* 3rd Ed., John Wiley & Sons, 1994.

Biomedical Instrumentation and Safety

Thomas R. Myers

INTRODUCTION

Medical instrumentation uses transducers, signal conditioners, and displays to
convert real-time information about living systems to a form that people can
interpret. Medical instrumentation is used for research, monitoring, and diagnosis
purposes. Ideally, the instrumentation will not alter the quantities being measured.

Medical instrumentation should also be designed to assure the safety of the subject and investigator.

SYSTEMS

The physical quantity that is measured is called the measurand. Most measurands fall into the following categories: biopotentials, pressures, movements, impedances, temperatures, and concentrations.

Some measurands change so slowly that they require only occasional sampling. Some rapidly changing measurands require continuous monitoring. The frequency content of the measurand, the objective of the measurement, the condition of the subject, and the liability of the investigator all influence data acquisition.

A transducer is a device that converts one form of energy to another. Ideally, a transducer will respond to only one form of energy, to the exclusion of all other forms of energy.

Often, measurands can be coupled directly to a transducer because the measurand is directly available. When the measurand is not directly available, another measurand that bears a known relationship to the desired measurand is utilized, or a transducer that interacts with the desired measurand to create an indirect measurand is utilized.

Usually transducer outputs cannot be directly fed to display devices. Signal conditioners often only amplify and filter transducer outputs, but also often compensate for adverse transducer characteristics, or average out noise.

The ideal form of display depends on the particular measurand, and the information sought by the investigator. Displays are usually numerical or graphical, digital or analog, temporary or permanent, real-time or stored and post-processed, visual or auditory. Displays are usually calibrated to a standard.

OUTPUT DISPLAYS

The result of the biopotential acquisition process is useless unless it is presented in a form that can be understood, interpreted, and acted on by the investigator. As with signal conditioners, displays should be easy to calibrate. In addition, during the data acquisition process, when time pressure and distractions can be significant, displays should present the data that the investigator is attempting to capture and isolate it in as clearly a recognizable format as is possible.

MEASURANDS

Many variables in biological systems are inaccessible because a good measurand to transducer interface does not exist. It is rarely possible to turn off biological systems or temporarily remove parts for measurement. Interference from other physiologic systems and physical aspects of transducers often prevent a good interface.

Variables measured from biologic systems are rarely completely deterministic. Most measurements vary over time and vary widely among subjects. These variations are often the result of interactions and feedback loops between many physiologic systems. Rarely is an investigator able to control or eliminate the effects of other physiologic systems on the desired variable. Often statistical methods are utilized to establish distribution functions of normal measurement variability.

Desired inputs are those measurands that the instrument is designed to isolate. Interfering inputs are those quantities that affect the instrument as an unwanted consequence of acquiring the desired measurands. Modifying inputs are those quantities that perform undesired modifications of instrument performance.

Biomedical measurements involving energy being transferred to a transducer or to living tissue must not exceed safe energy levels as reversible physiologic changes may affect measurements. Biomedical instruments must be designed to maintain the safety of both the patient and the investigator.

The following table lists many common biomedical measurements and their typical ranges, bandwidth, and transducers (from C. D. Ray, *Medical Engineering*, Year Book Medical Publishers, 1974).

Electrocardiography

The heart is the four-chambered pump of the circulatory system. The primary pumping action is performed by the two ventricle chambers, while the two atrial antechambers serve to collect blood during the time the ventricles are pumping.

Table 9.1 Common biomedical measurements

Measurand	Range	Bandwidth	Transducer
Ballistocardiography	0.7 mG	0.40 Hz	Accelerometer
Ballistocardiography	0–100 μm	0–40 Hz	Displacement
Bladder Pressure	1–100 cm H_2O	0–10 Hz	Strain Gauge Manometer
Blood Flow	1–300 ml/sec	0–20 Hz	Flowmeter
Arterial Blood Pressure	10–400 mm Hg	0–50 Hz	Manometer, Direct
Arterial Blood Pressure	25–400 mm Hg	0–60 Hz	Cuff, Indirect
Venous Blood Pressure	0–50 mm Hg	0–50 Hz	Manometer, Indirect
Blood Gas, O_2	30–100 mm Hg	0–2 Hz	Specific Electrode
Blood Gas, CO_2	40–100 mm Hg	0–2 Hz	Specific Electrode
Blood Gas, N_2	1–3 mm Hg	0–2 Hz	Specific Electrode
Blood Gas, CO	0.1–0.4 mm Hg	0–2 Hz	Specific Electrode
Blood pH	6.7–7.8 pH	0–2 Hz	Specific Electrode
Cardiac Output	4–25 l/min	0–20 Hz	Flowmeter
Electrocardiography	0.5–4 mV	0.01–250 Hz	Skin Electrodes
Electroencephalography	5–300 μV	0–150 Hz	Scalp Electrodes
Electrocorticography	10–5000 μV	0–150 Hz	Brain Surface Electrodes
Electrogastrography	10–1000 μV	0–1 Hz	Skin, Stomach Electrodes
Electromyography	0.1–5 mV	0–10 kHz	Needle Electrodes
Electroocculography	50–3500 μV	0–50 Hz	Contact Electrodes
Electroretinography	0–900 μV	0–50 Hz	Contact Electrodes
Galvanic Skin Response	1–500 KΩ	0.01–1 Hz	Skin Electrodes
Gastric pH	3–13 pH	0–1 Hz	pH Electrode
Gastrointestinal Pressure	0–100 cm H_2O	0–10 Hz	Strain Gauge Manometer
Gastrointestinal Forces	0–50 g	0–1 Hz	Displacement
Nerve Potentials	0.01–3 mV	0–10 kHz	Skin, Needle Electrodes
Phonocardiography	80 dB	5–2 kHz	Microphone
Plethysmography, Volume	Organ Dependent	0–30 Hz	Displacement, Impedance
Plethysmography, Circulation	Organ Dependent	0–30 Hz	Displacement, Impedance
Pneumotachography	0–600 l/min	0–40 Hz	Pressure
Respiratory Rate	2–50 /min	0.1–10 Hz	Impedance, Thermistor
Tidal Volume	50–1000 ml	0.1–10 Hz	Impedance, Thermistor
Temperature	32–40°C	0–0.1 Hz	Thermocouple

The rhythmic contraction of the chambers is reflected in the simultaneous set of electrical events that occur within the heart. This set of electrical events is intrinsic to the action of the heart. The coordinated contraction of the atria and the ventricles is orchestrated by the specific pattern of electrical activation potentials within the muscles of the heart. The electrical activation patterns in the atria and ventricles are triggered by the coordinated series of electrical events from within the synchronous conduction system of the heart.

The electrical activation sequence of the heart leads to the flow of closed path currents within the cardiac volume conductor. The potentials measured at the outer body surface as a result of these currents are referred to as the electrocardiogram, or ECG.

Three primary skin-mounted leads make up the frontal plane ECG. The right arm, left arm, and left leg electrodes make up the so-called Eindhoven's Triangle. Also, a reference electrode is usually placed on the right leg.

Electroretinography

The retina, located on the rear inner surface of the eye, is the sensory element of the eye. Light passes through the cornea, the anterior chamber, the lens, and then the vitreous chamber before striking the retina. The anterior chamber is filled with a transparent liquid, the aqueous humor. The vitreous chamber is filled with a transparent gel, the vitreous body.

When the retina is stimulated by light, a sequence of electrical events can be sensed between an exploring electrode, placed at either the retina or cornea, and a remote reference electrode, usually placed at either the earlobe or forehead. The electrical event sequence is known as the electroretinogram, or ERG.

The ERG is obtained from an Ag-AgCl exploring electrode mounted within a saline filled contact lens. Like the ECG, the ERG is an externally sensed wave form that occurs as a byproduct of complex internal electrical events.

In addition to the transient potential ERG, there is a steady state corneal-retinal potential known as the electrooculogram, or EOG. The EOG is a function of rotational eye position, and is sensed with skin electrodes placed beside the eye. There is an almost linear relationship between the angle of gaze and EOG output up to about 30° away from straight ahead.

Electroencephalography

The term electroencephalogram, or EEG, denotes the potentials recorded from the brain using scalp electrodes. The term electrocorticogram, or ECOG, denotes the potentials recorded from the brain using cortical surface or needle depth electrodes.

The sensed potential fluctuation represents the summation of the volume conductor fields produced by enormous collection of the brain's neuronal current generators. The sources generating the potentials recorded here are the aggregates of complex central nervous system elements and interconnections.

Electroneurography

Though the field potentials from the peripheral nerves are of much smaller amplitude than the field potentials from the surrounding excitable muscle fibers, the neural potentials can nonetheless be recorded with either needle electrodes or skin electrodes. Nerve field potentials can be isolated by applying stimuli in a manner that does not evoke a muscle response.

The electroneurogram, or ENG, is often used to determine the conduction velocity of a peripheral nerve by stimulating or recording from the nerve at two different points a known distance apart, and measuring the conduction time difference between the two points.

Electromyography

The motor unit is the smallest group of skeletal muscle fibers that may be activated synchronously by volition. The evoked field potential from the active fibers of a motor unit is called an electromyogram, or EMG, and may be sensed most conveniently with skin electrodes. However, the disadvantages of using skin electrodes are that the technique only works with superficial muscles, and the electrodes are sensitive to electrical activity over a wide area.

Electromagnetic Blood Flowmeter

When an electrical conductor moves through a magnetic field, cutting across the lines of magnetic flux, a potential is induced along the length of the conductor. Blood is a conductor, having approximately the same conductance as saline. The electromagnetic flowmeter creates a uniform magnetic field across a blood vessel, and measures the induced potential over the length of the blood vessel in the field due to the flow of the conductive blood. The induced potential is directly proportional to the velocity of the blood flowing within the vessel.

The electromagnetic flowmeter is sensitive enough to measure the instantaneous pulsatile flow of blood.

Ultrasonic Blood Flowmeter

The ultrasonic blood flowmeter utilizes piezoelectric crystal transducers to inject acoustic energy into a blood vessel, and to sense acoustic energy reflected back from the blood vessel. The frequency of the acoustic energy is shifted in proportion to the velocity of the blood flowing through the vessel.

Blood Flow Plethysmography

Plethysmography is the measurement of the volume or the change in volume of a portion of the body. Venous occlusion plethysmography, an example of chamber plethysmography, uses a constrictive cuff to prevent venous blood from leaving a limb, and thus measures blood flow by measuring limb volume changes over time.

Capacitance plethysmography measures the capacitance between the skin of a limb and a cylindrical outer electrode. Changes in the blood volume of a limb are measured as capacitance changes between the skin and the outer cylinder.

Electrical impedance plethysmography measures the impedance across a segment of tissue. As the volume of the tissue changes due to pulsations of blood flowing through the tissue, the impedance across the tissue changes, and is measured.

Respiratory Plethysmography

Respiratory plethysmography is usually performed by either transthoracic impedance plethysmography, or by total-body plethysmography. As the volume of the lungs changes due to changes in the volume of air within the lungs, the electrical impedance across the lungs changes, and is measured as changes in the transthoracic impedance.

Total-body plethysmography is a form of chamber plethysmography in which the entire subject is enclosed in a rigid container, and pressure changes, volume displacement changes, and volume displacement rate changes are all monitored.

Photo Plethysmography

Light may be transmitted through a capillary bed, such as in a fingertip or an ear lobe. As blood pulsations flow through the capillary bed, changes in the volume of the capillaries change the absorbed, reflected, and transmitted quantities of the light as it passes through, which are detected by photo sensors. Heart rate is often measured by this means.

ELECTRODES

Typical biomedical signal transducers are electrodes of some form. The output impedance of the electrodes is usually very high, often in the hundreds of kilohms to tens of megohms range. The high impedance is due to the fact that the signal source often involves the flow of ions and electrons across semipermeable membranes and electrolytic to metallic interfaces. An undesirable side effect that results from these interfaces is a half cell potential in series with the output signal.

There are many techniques utilized to minimize the source impedance and half cell potential. These techniques include the use of subcutaneous skin electrodes, conductive electrode pastes, nonpolarizable electrodes constructed of noble metals (*e.g.*, Pt, Au, Ag), and electrochemical electrodes with an ionic mobility well matched to that of the source medium.

It is normally desired to construct electrodes from one type of metal to avoid any additional half cell potentials that would result from the joining of dissimilar metals. Common practice is also to obtain signals differentially from identical electrode pairs so that the electrode half cell potentials may be rejected as common mode signals. Another benefit of differential electrodes is the rejection of contamination from 60 Hz power line voltages for which biologic sources seem to act as perfect antennas.

SIGNAL CONDITIONERS

The primary function of a biopotential signal conditioner is to transform a weak and/or noisy electrical signal of biologic origin and amplify and/or filter it so that it may be further analyzed, saved, and displayed. Signal conditioners must minimize the loading of transducer outputs, in order to minimize distortion of the transducer output signals. Differential input conditioners must have high common mode rejection ratios and ranges in order to minimize artifacts related to common mode signals. In addition, signal conditioners should be easy to calibrate.

Often, signal conditioners electrically isolate their output from their inputs to cut ground loops, isolate transducers from noise sources, and/or to isolate subjects from potential shock sources. Any current or potential that appears across the signal conditioner inputs must be incapable of affecting either the subject or the bipotential being measured.

Signal conditioners usually filter out any contaminating signals, such as noise, or other biopotentials that are detected by the transducer. Optimal signal-to-noise ratios are obtained by adjusting the pass bandwidth of the conditioner to be just wide enough to process the desired measurand without distortion, but not wide enough to pass undesired noise. Unwanted biopotentials are often also removed by filtering, but more complex techniques such as time gating, averaging, adaptive filtering, and selective processing are also utilized.

INSTRUMENTATION AMPLIFIERS

Instrumentation amplifiers are regularly utilized as signal conditioners because they offer differential high impedance inputs, high differential gain, low common mode gain, and thus high common mode rejection ratio (CMRR). The schematic of a widely utilized instrumentation amplifier is shown in Figure 9.1. The input impedance of the noninverting V^+ input and the inverting V^- input is the input impedance of the op-amps, which is as high as about 10^{12} ohms with state-of-the-art commercially available op-amps.

The differential gain G_D is defined as the amplifier gain for $V^+ \neq V^-$. The common mode gain G_{CM} is defined as the amplifier gain for $V^+ = V^-$. The CMRR is defined as the ratio G_D/G_{CM}.

The gains are derived as follows:

V_a is virtually equal to V^-;

V_b is virtually equal to V^+;

so $(V_d - V_c)/(R_1 + R_2 + R_3) = (V^+ - V^-)/R_2$;

V_e is virtually equal to V_f;

So $V_d \cdot R_7/(R_6 + R_7) = V_o + ((V_c - V_o) \cdot R_5/(R_4 + R_5))$.

Using closely matched resistors so that $R_3 \approx R_1$, and $R_6 \approx R_4$, and $R_7 \approx R_5$,

$$G_D \approx (2R_1 + R_2) \cdot R_5/(R_2 \cdot R_4);$$
$$G_{CM} \approx [(R_5/(R_5 + R_4)) - (R_7/(R_7 + R_6))]/(R_4/(R_5 + R_4));$$
$$\text{CMRR} \approx (2R_1 + R_2) \cdot (R_5/(R_5 + R_4))/[(R_5/(R_5 + R_4)) - (R_7/(R_7 + R_6))] \cdot R_2.$$

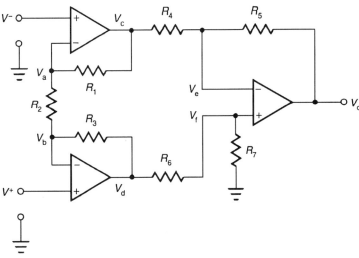

Figure 9.1 Typical instrumentation amplifier

Using exactly matched resistors so that $R_3 = R_1$, and $R_6 = R_4$, and $R_7 = R_5$,

$$G_D = (2R_1 + R_2) \cdot R_5/(R_2\,R_4);$$
$$G_{CM} = 0;$$
$$\text{CMRR} = \infty.$$

In reality, non-infinite op-amp gains nonzero op-amp input bias currents and offset voltages, and imperfectly matched resistor pairs conspire to yield a finite CMRR. A CMRR of at least 100 dB is achievable with the amplifier.

An Example Design Question

For the instrumentation amplifier just described, what are G_D, G_{CM}, and CMRR for $R_2 = 10$ kΩ, and the following 0.1% matched resistor pairs: $R_1 = 45.00$ kΩ, $R_3 = 45.00$ kΩ; $R_4 = 10.00$ kΩ, $R_6 = 10.01$ kΩ; $R_5 = 1.00$ MΩ, $R_7 = 999.9$ kΩ? Assume ideal op-amps.

$$G_D = 1000 = 60.0 \text{ dB};$$
$$G_{CM} = 0.001089 = -59.3 \text{ dB};$$
$$\text{CMRR} = 918100 = 119.3 \text{ dB}.$$

LINEAR ISOLATION AMPLIFIERS

As the name implies, linear isolation amplifiers pass signals from their inputs to their outputs without possessing an electrical path between their inputs and their outputs. The passed signals are not distorted by the amplifier along the way. The input side of the amplifier is electrically isolated from the output side of the amplifier. Isolation is generally achieved with one of two methods: magnetic isolation or optical isolation.

Magnetic isolators modulate an oscillator with the input signal, pass the modulated signal through a transformer, demodulate the transformer output signal, and pass the demodulated signal to the output.

Optical isolators convert the input signal energy to light energy, typically with a light emitting diode. The light energy is then converted back to electrical energy, typically with a photosensitive transistor, and then passed to the output.

Both types of isolators often employ isolated feedback loops back across the isolation barrier in order to linearize the net effect of the energy conversion processes. Highly linear isolators with barrier impedances greater than $10^{12}\Omega$ are commercially available.

Power for the input side of isolation amplifiers is provided by isolated power supplies that employ transformers in much the same manner as do magnetic linear isolation amplifiers.

There are several reasons why isolation amplifiers are employed in the signal conditioning process. Isolation amplifiers separate the patient circuits from the signal conditioning circuits, thus eliminating paths for electrical noise to proceed from the patient to the signal conditioner. Isolation amplifiers break potential signal and ground loops involving both the patient and the signal conditioner circuits. Such loops can act as antennas for unwanted noise, especially of the 60 Hz power line variety. Most importantly, isolation amplifiers break potential shock paths between the patient and the equipment to which he or she is connected.

SHOCKS

In general, shocks occur when the patient accidentally becomes part of a circuit between a 60 Hz power line source and ground. The shock circuit is often quite roundabout and nonobvious, which probably has a great deal to do with why one is occasionally formed. The physiological damage that results from shock takes the form of resistive heating and electrochemical burning of the tissue, and the stimulation of excitable nerve and muscle tissue.

Macroshocks

When shock current enters and exits the body through the skin, generally only a small fraction of the current flows through the heart. In order for current levels through the heart to reach dangerous levels, the externally applied currents must be rather large, and are thus referred to as macroshocks.

The following macroshock effects are the result of 60 Hz currents applied to the hands. The wideness of the current ranges of patient susceptibilities described is due to the wideness of the ranges of local current densities that result in different patients at different current levels.

The threshold of perception occurs when enough current flows to excite the nerve endings of the skin. The range of threshold of perception current is from about $^1/_2$ milliamp to about 7 milliamps.

Involuntary muscle contractions result from higher current levels. The maximum "let go" current is defined as the highest current level that a subject can voluntarily withdraw from. The range of maximum "let go" current is from about 7 milliamps to about 100 milliamps.

Involuntary contraction and paralysis of the respiratory muscles results from even higher current levels. The range of minimum respiratory arrest current is from about 10 milliamps to about 22 milliamps.

At even higher current levels, if the density of current through only part of the heart is sufficient enough to excite cardiac muscle, then the normal sequence of synchronizing electrical events within the heart can be disrupted. Once the heart has been desynchronized, or fibrillated, it stops pumping. Simply ceasing the flow of the shock current will not start the heart pumping again. The range of the threshold of fibrillation current is from about 75 milliamps to about 400 milliamps.

Still higher current levels are capable of simultaneously depolarizing the entire heart muscle instead of just part of it. This concurrent relaxation, or defibrillation, results in the return of the heart's normal rhythmic activity. The range of the maximum fibrillation or minimum defibrillation current is about 1 amp to about 6 amps.

Yet higher current levels result in sustained contraction of the entire heart muscle. Normal heart activity ensues when the shock current is ceased. The range of the minimum complete myocardial contraction current is about 1 amp to about 6 amps.

Above the 1 amp to 6 amps current range, burns due to resistive heating usually result. Burns occur particularly at the skin contact points because of the high resistance of skin as compared to the lower resistance of internal tissues. At these high current levels, nervous tissue loses functional excitability, and skeletal muscles undergo severe contraction.

Macroshock hazards in the biomedical environment generally come from equipment with faulty hot and/or ground power line connections. A 1 Ω resistance ground connection between equipment chassis ground and earth ground will conduct virtually all of the current to earth ground from a hot power line conductor with faulty insulation that is contacting chassis ground. Virtually no current flows to earth ground through a high resistance person touching the chassis.

However, were the chassis ground to earth ground connection to fail as an open circuit, a macroshock hazard would be present. A person simultaneously touching the hot chassis and any other earth grounded object would then conduct all of the current from the faulty hot power line conductor to earth ground.

Microshocks

When shock current enters and exits the body directly through the heart, much lower current levels are considered dangerous. These internally applied currents are thus referred to as microshocks. Patients that have direct electrical connections to the heart are termed electrically susceptible patients.

Data for fibrillation caused by shock currents passing through intracardiac catheters indicate that the range of the minimum fibrillation current is from about 80 microamps to about 600 microamps. The safety current limit for microshocks is 10 microamps.

Microshock hazards in the biomedical environment generally come from leakage currents that invariably flow between insulated conductors at different potentials. Most of the leakage current flows through capacitive coupling between conductors, though some resistive leakage current flows through insulation, dirt, and moisture.

The leakage currents pertinent to microshocks are those leakage currents that ultimately flow: between conductors with direct connections to the patient and line powered conductors; and between conductors with direct connections to the patient and chassis ground conductors.

A microshock hazard is present for a patient who has a heart catheter connected to a piece of equipment with excessive capacitive coupling between the conductive fluid in the catheter and the power supply lines of the equipment. Were the patient to touch the earth grounded hospital bed frame, a microshock circuit would be completed.

Microshock hazards in the biomedical environment can also come from current flowing through a patient connected between instruments with a voltage difference between the different chassis ground potentials.

A microshock hazard is present for a patient with an ECG monitor plugged into one outlet, and a heart catheter blood pressure monitor plugged into a different outlet. Any ground fault current flowing through the ground line between the two outlets would place the chassis of the two instruments at different potentials. Capacitive coupling between the ECG leads and ECG chassis ground, and between the catheter and the book pressure chassis ground, would complete a circuit through which a microshock current could flow.

The schematic of a microshock circuit is shown in Figure 9.2.

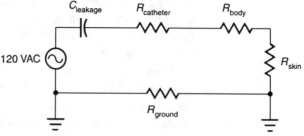

Figure 9.2 A microshock circuit

An Example Microshock Question

A patient in the cardiac care unit is lying in a bed with its frame grounded to the hospital's electrical ground network. The series resistance of a hospital's electrical ground network is usually negligible (less than 1Ω). The patient's hand is resting on the bed frame. The series resistance of a patient's skin is about $100k\Omega$. The internal series resistance between a patient's hand and heart is about 300Ω. A saline-filled catheter is connected between a patient's heart and a 120 VAC line-powered blood pressure monitor. The series resistance of the saline between a patient's heart and the monitor's pressure transducer is about $40\,k\Omega$. The blood pressure monitor has leaky insulation in its power supply. The capacitance between the hot lead of the line cord and the leads of the pressure transducer is about $0.063\,\mu F$.

Is the patient described above in danger of fibrillation? At 60 Hz, the series impedance of the leakage capacitor is $Z = 1/(2 \cdot \pi \cdot 60 \text{ Hz} \cdot 0.063\,\mu F) \approx 100\,k\Omega$. The total series resistance is $R \approx 240\,k\Omega$. The microshock current $I \approx 120V/240\,k\Omega = 500\,\mu A$. The patient is in great danger of fibrillation as the safety current limit is $10\,\mu A$.

Shock Protection and Prevention

The typical macroshock circuit path runs from the utility company hot power line, through a person to earth ground, and through earth ground to the grounding point of the utility company neutral power line. Installing a 1:1 input-to-output ratio isolation transformer into the power line path is one method of protecting against macroshocks. The transformer's primary coil is connected to the utility company's hot and neutral power lines. The transformer's secondary coil provides the isolated hot and neutral power lines. The macroshock circuit path has been broken by the insertion of the isolation transformer because there is now no electrical path between the utility company's power line and the isolated hot power line.

The microshock circuit path has unfortunately not been broken by the insertion of the isolation transformer because there are capacitive and resistive leakage current paths between the input and output coils of the transformer. Though useful for protecting against macroshocks, the leakage currents through isolation transformers are too high for the transformers to be used alone to protect against microshocks.

Installing ground fault circuit interrupters (GFCIs) in series with the power lines is another method of protecting against macroshocks. Any imbalance over a few milliamps between the hot and neutral power lines indicates that current is flowing to ground through an undesired path, and the interrupter cuts off power to its output.

However, GFCIs should not be used to protect critical life-support equipment. In this case, the tripping off of the power is probably more life-threatening to a patient than a ground fault. The minimum trip current of GFCIs is generally too high for the interrupters to be used alone to protect against microshocks.

Microshock protection is achieved through complete electrical isolation of the patient. All electrical signals are passed through the types of linear isolation amplifiers and isolated power supplies described previously. Electrical power is provided through isolation transformers. Catheters for electrified equipment have their transducers located at the patient's body, and the electrical connections to the transducer are, as above, isolated. Electrified equipment utilizes low-voltage power supplies, and is constructed with low leakage insulation. Double-insulated construction is also utilized that prevents any contact with the grounded chassis of equipment.

Equipotential grounding systems are also important to microshock protection. By grounding all equipment within the patient environment to a single grounding point, potential differences between equipment chassis grounds due to current flow through ground loops can be avoided.

The maintenance of all the shock elimination means described above is accomplished through the regular testing of equipment for low leakage currents, low resistance of interconnections, and the integrity of insulation.

RECOMMENDED REFERENCE

C.D. Ray, *Medical Engineering,* Year Book Medical Publishers, 1974.

Engineering Economics

Donald G. Newnan

OUTLINE

This is a review of the field known variously as *engineering economics, engineering economy*, or *engineering economic analysis*. Since engineering economics is straightforward and logical, even people who have not had a formal course should be able to gain sufficient knowledge from this chapter to successfully solve most engineering economics problems.

There are 35 example problems scattered throughout the chapter. These examples are an integral part of the review and should be examined as you come to them.

The field of engineering economics uses mathematical and economics techniques to systematically analyze situations which pose alternative courses of action. The initial step in engineering economics problems is to resolve a situation, or each alternative in a given situation, into its favorable and unfavorable consequences or factors. These are then measured in some common unit—usually money. Factors which cannot readily be equated to money are called intangible or irreducible factors. Such factors are considered in conjunction with the monetary analysis when making the final decision on proposed courses of action.

CASH FLOW

A cash flow table shows the "money consequences" of a situation and its timing. For example, a simple problem might be to list the year-by-year consequences of purchasing and owning a used car:

Year	Cash Flow	
Beginning of first year 0	−$4500	Car purchased "now" for $4500 cash. The minus sign indicates a disbursement.
End of year 1	−350	
End of year 2	−350	
End of year 3	−350	Maintenance costs are $350 per year.
End of year 4	−350	
	+2000	This car is sold at the end of the fourth year for $2000. The plus sign represents the receipt of money.

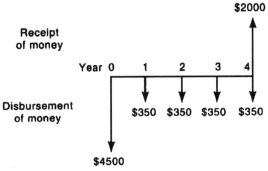

Figure A.1

This same cash flow may be represented graphically, as shown in Figure A.1. The upward arrow represents a receipt of money, and the downward arrows represent disbursements. The horizontal axis represents the passage of time.

| Example **A.1** | In January 1993 a firm purchased a used typewriter for $500. Repairs cost nothing in 1993 or 1994. Repairs are $85 in 1995, $130 in 1996, and $140 in 1997. The machine is sold in 1997 for $300. Complete the cash flow table. |

Solution

Unless otherwise stated, the customary assumption is a beginning-of-year purchase, followed by end-of-year receipts or disbursements, and an end-of-year resale or salvage value. Thus the typewriter repairs and the typewriter sale are assumed to occur at the end of the year. Letting a minus sign represent a disbursement of money and a plus sign a receipt of money, we are able to set up the cash flow table:

Year	Cash Flow
Beginning of 1993	−$500
End of 1993	0
End of 1994	0
End of 1995	−85
End of 1996	−130
End of 1997	+160

Notice that at the end of 1997 the cash flow table shows +160, which is the net sum of −140 and +300. If we define year 0 as the beginning of 1993, the cash flow table becomes

Year	Cash Flow
0	−$500
1	0
2	0
3	−85
4	−130
5	+160

From this cash flow table, the definitions of year 0 and year 1 become clear. Year 0 is defined as the *beginning* of year 1. Year 1 is the *end* of year 1, and so forth.

TIME VALUE OF MONEY

When the money consequences of an alternative occur in a short period of time—say, less than one year—we might simply add up the various sums of money and obtain the net result. But we cannot treat money this way over longer periods of time. This is because money today does not have the same value as money at some future time.

Consider this question: Which would you prefer, $100 today or the assurance of receiving $100 a year from now? Clearly, you would prefer the $100 today. If you had the money today, rather than a year from now, you could use it for the year. And if you had no use for it, you could lend it to someone who would pay interest for the privilege of using your money for the year.

Simple Interest

Simple interest is interest that is computed on the original sum. Thus if one were to lend a present sum P to someone at a simple annual interest rate i, the future amount F due at the end of n years would be

$$F = P + Pin$$

Example **A.2**

How much will you receive back from a $500 loan to a friend for three years at 10 percent simple annual interest?

Solution

$$F = P + Pin = 500 + 500 \times 0.10 \times 3 = \$650$$

In Example A.2 one observes that the amount owed, based on 10 percent simple interest at the end of one year, is $500 + 500 \times 0.10 \times 1 = \$550$. But at simple interest there is no interest charged on the $50 interest, even though it is not paid until the end of the third year. Thus simple interest is not realistic and is seldom used. *Compound interest* charges interest on the principal owed plus the interest earned to date. This produces a charge of interest on interest, or compound interest. Engineering economics uses compound interest computations.

EQUIVALENCE

In the preceding section we saw that money at different points in time (for example, $100 today or $100 one year hence) may be equal in the sense that they both are $100, but $100 a year hence is *not* an acceptable substitute for $100 today. When we have acceptable substitutes, we say they are *equivalent* to each other. Thus at 8 percent interest, $108 a year hence is equivalent to $100 today.

Example **A.3**

At a 10 percent per year (compound) interest rate, $500 now is *equivalent* to how much three years hence?

Solution

A value of $500 now will increase by 10 percent in each of the three years.

$$Now = \$500.00$$
$$\text{End of 1st year} = 500 + 10\%(500) = 550.00$$
$$\text{End of 2nd year} = 550 + 10\%(550) = 605.00$$
$$\text{End of 3rd year} = 605 + 10\%(605) = 665.50$$

Thus $500 now is *equivalent* to $665.50 at the end of three years. Note that interest is charged each year on the original $500 plus the unpaid interest. This compound interest computation gives an answer that is $15.50 higher than the simple-interest computation in Example A.2.

Equivalence is an essential factor in engineering economics. Suppose we wish to select the better of two alternatives. First, we must compute their cash flows. For example,

	Alternative	
Year	*A*	*B*
0	−$2000	−$2800
1	+800	+1100
2	+800	+1100
3	+800	+1100

The larger investment in alternative *B* results in larger subsequent benefits, but we have no direct way of knowing whether it is better than alternative *A*. So we do not know which to select. To make a decision, we must resolve the alternatives into *equivalent* sums so that they may be compared accurately.

COMPOUND INTEREST

To facilitate equivalence computations, a series of compound interest factors will be derived here, and their use will be illustrated in examples.

Symbols and Functional Notation

i = effective interest rate per interest period. In equations, the interest rate is stated as a decimal (that is, 8 percent interest is 0.08).

n = number of interest periods. Usually the interest period is one year, but it could be something else.

P = a present sum of money.

F = a future sum of money. The future sum F is an amount n interest periods from the present that is equivalent to P at interest rate i.

A = an end-of-period cash receipt or disbursement in a uniform series continuing for n periods. The entire series is equivalent to P or F at interest rate i.

Table A.1 Periodic compounding: Functional notation and formulas

Factor	Given	To Find	Functional Notation	Formula
Single payment				
Compound amount factor	P	F	$(F/P, i\%, n)$	$F = P(1 + i)^n$
Present worth factor	F	P	$(P/F, i\%, n)$	$P = F(1 + i)^{-n}$
Uniform payment series				
Sinking fund factor	F	A	$(A/F, i\%, n)$	$A = F\left[\dfrac{i}{(1+i)^n - 1}\right]$
Capital recovery factor	P	A	$(A/P, i\%, n)$	$A = P\left[\dfrac{i(1+i)^n}{(1+i)^n - 1}\right]$
Compound amount factor	A	F	$(F/A, i\%, n)$	$F = A\left[\dfrac{(1+i)^n - 1}{i}\right]$
Present worth factor	A	P	$(P/A, i\%, n)$	$P = A\left[\dfrac{(1+i)^n - 1}{i(1+i)^n}\right]$
Uniform gradient				
Gradient present worth	G	P	$(P/G, i\%, n)$	$P = G\left[\dfrac{(1+i)^n - 1}{i^2(1+i)^n} - \dfrac{n}{i(1+i)^n}\right]$
Gradient future worth	G	F	$(F/G, i\%, n)$	$F = G\left[\dfrac{(1+i)^n - 1}{i^2} - \dfrac{n}{1}\right]$
Gradient uniform series	G	A	$(A/G, i\%, n)$	$A = G\left[\dfrac{1}{i} - \dfrac{n}{(1+i)^n - 1}\right]$

G = uniform period-by-period increase in cash flows; the uniform gradient.

r = nominal annual interest rate.

From Table A.1 we can see that the functional notation scheme is based on writing (to find/given, i, n). Thus, if we wished to find the future sum F, given a uniform series of receipts A, the proper compound interest factor to use would be $(F/A, i, n)$.

Single-Payment Formulas

Suppose a present sum of money P is invested for one year at interest rate i. At the end of the year, the initial investment P is received together with interest equal to Pi, or a total amount $P + Pi$. Factoring P, the sum at the end of one year is $P(1 + i)$. If the investment is allowed to remain for subsequent years, the progression is as follows:

Amount at Beginning of the Period	+	Interest for the Period	=	Amount at End of the Period
1st year, P	+	Pi	=	$P(1 + i)$
2nd year, $P(1 + i)$	+	$Pi(1 + i)$	=	$P(1 + i)^2$
3rd year, $P(1 + i)^2$	+	$Pi(1 + i)^2$	=	$P(1 + i)^3$
nth year, $P(1 + i)^{n-1}$	+	$Pi(1 + i)^{n-1}$	=	$P(1 + i)^n$

The present sum P increases in n periods to $P(1 + i)^n$. This gives a relation between a present sum P and its equivalent future sum F:

$$\text{Future sum} = (\text{present sum})(1 + i)^n$$
$$F = P(1 + i)^n$$

This is the *single-payment compound amount formula*. In functional notation it is written

$$F = P(F/P, i, n)$$

The relationship may be rewritten as

$$\text{Present sum} = (\text{Future sum}) (1 + i)^{-n}$$
$$P = F(1 + i)^{-n}$$

This is the *single-payment present worth formula*. It is written

$$P = F(P/F, i, n)$$

Example A.4

At a 10 percent per year interest rate, $500 now is *equivalent* to how much three years hence?

Solution

This problem was solved in Example A.3. Now it can be solved using a single-payment formula. $P = \$500$, $n = 3$ years, $i = 10$ percent, and $F =$ unknown:

$$F = P(1 + i)^n = 500(1 + 0.10)^3 = \$665.50.$$

This problem also may be solved using a compound interest table:

$$F = P(F/P, i, n) = 500(F/P, 10\%, 3)$$

From the 10 percent compound interest table, read $(F/P, 10\%, 3) = 1.331$.

$$F = 500(F/P, 10\%, 3) = 500(1.331) = \$665.50$$

Example A.5

To raise money for a new business, a man asks you to lend him some money. He offers to pay you $3000 at the end of four years. How much should you give him now if you want 12 percent interest per year?

Solution

$P =$ unknown, $F = \$3000$, $n = 4$ years, and $i = 12$ percent:

$$P = F(1 + i)^{-n} = 3000(1 + 0.12)^{-4} = \$1906.55$$

Alternative computation using a compound interest table:

$$P = F(P/F, i, n) = 3000(P/F, 12\%, 4) = 3000(0.6355) = \$1906.50$$

Note that the solution based on the compound interest table is slightly different from the exact solution using a hand-held calculator. In engineering economics the compound interest tables are always considered to be sufficiently accurate.

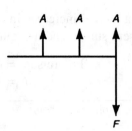

A = End-of-period cash receipt or disbursement in a uniform series continuing for n periods

F = A future sum of money

Figure A.2

Uniform Payment Series Formulas

Consider the situation shown in Figure A.2. Using the single-payment compound amount factor, we can write an equation for F in terms of A:

$$F = A + A(1 + i) + A(1 + i)^2 \qquad \text{(i)}$$

In this situation, with $n = 3$, Eq. (i) may be written in a more general form:

$$F = A + A(1 + i) + A(1 + i)^{n-1} \qquad \text{(ii)}$$

Multiply Eq. (ii) by $(1 + i)$ $\quad (1 + i)F = A(1 + i) + A(1 + i)^{n-1} + A(1 + i)^n \qquad \text{(iii)}$

Subtract Eq. (ii): $\qquad\qquad -F = A + A(1 + i) + A(1 + i)^{n-1} \qquad \text{(ii)}$

$$iF = -A + A(1 + i)^n$$

This produces the *uniform series compound amount formula:*

$$F = A\left(\frac{(1+i)^n - 1}{i}\right)$$

Solving this equation for A produces the *uniform series sinking fund formula:*

$$A = F\left(\frac{i}{(1+i)^n - 1}\right)$$

Since $F = P(1 + i)^n$, we can substitute this expression for F in the equation and obtain the *uniform series capital recovery formula:*

$$A = P\left(\frac{i(1+i)^n}{(1+i)^n - 1}\right)$$

Solving the equation for P produces the *uniform series present worth formula:*

$$P = A\left(\frac{(1+i)^n - 1}{i(1+i)^n}\right)$$

In functional notation, the uniform series factors are

Compound amount $(F/A, i, n)$

Sinking fund $(A/F, i, n)$

Capital recovery $(A/P, i, n)$

Present worth $(P/A, i, n)$

Example **A.6**

If $100 is deposited at the end of each year in a savings account that pays 6 percent interest per year, how much will be in the account at the end of five years?

Solution

A = $100, F = unknown, n = 5 years, and i = 6 percent:

$$F = A(F/A, i, n) = 100(F/A, 6\%, 5) = 100(5.637) = \$563.70$$

Example **A.7**

A fund established to produce a desired amount at the end of a given period, by means of a series of payments throughout the period, is called a *sinking fund*. A sinking fund is to be established to accumulate money to replace a $10,000 machine. If the machine is to be replaced at the end of 12 years, how much should be deposited in the sinking fund each year? Assume the fund earns 10 percent annual interest.

Solution

Annual sinking fund deposit A = 10,000(A/F, 10%, 12)

$$= 10,000(0.0468) = \$468$$

Example **A.8**

An individual is considering the purchase of a used automobile. The total price is $6200. With $1240 as a down payment, and the balance paid in 48 equal monthly payments with interest at 1 percent per month, compute the monthly payment. The payments are due at the end of each month.

Solution

The amount to be repaid by the 48 monthly payments is the cost of the automobile *minus* the $1240 downpayment.

P = $4960, A = unknown, n = 48 monthly payments, and i = 1 percent per month:

$$A = P(A/P, 1\%, 48) = 4960(0.0263) = \$130.45$$

Example **A.9**

A couple sell their home. In addition to cash, they take a mortgage on the house. The mortgage will be paid off by monthly payments of $450 for 50 months. The couple decides to sell the mortgage to a local bank. The bank will buy the mortgage, but it requires a 1 percent per month interest rate on its investment. How much will the bank pay for the mortgage?

Solution

A = $450, n = 50 months, i = 1 percent per month, and P = unknown:

$$P = A(P/A, i, n) = 450(P/A, 1\%, 50) = 450(39.196) = \$17,638.20$$

Uniform Gradient

At times one will encounter a situation where the cash flow series is not a constant amount A. Instead, it is an increasing series. The cash flow shown in Figure A.3 may

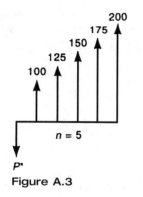

Figure A.3

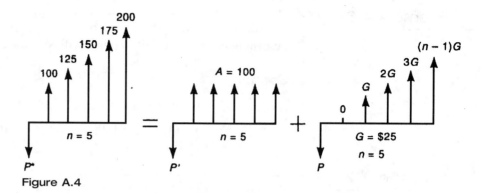

Figure A.4

be resolved into two components (Figure A.4). We can compute the value of P^* as equal to P' plus P. And we already have the equation for P': $P' = A(P/A, i, n)$. The value for P in the right-hand diagram is

$$P = G\left[\frac{(1+i)^n - 1}{i^2(1+i)^n} - \frac{n}{i(1+i)^n}\right]$$

This is the *uniform gradient present worth formula.* In functional notation, the relationship is $P = G(P/G, i, n)$.

Example A.10

The maintenance on a machine is expected to be $155 at the end of the first year, and it is expected to increase $35 each year for the following seven years (Exhibit 1). What sum of money should be set aside now to pay the maintenance for the eight-year period? Assume 6 percent interest.

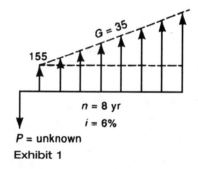

Exhibit 1

Solution

$$P = 155(P/A, 6\%, 8) + 35(P/G, 6\%, 8)$$
$$= 155(6.210) + 35(19.841) = \$1656.99$$

In the gradient series, if—instead of the present sum, P—an equivalent uniform series A is desired, the problem might appear as shown in Figure A.5. The relationship between A' and G in the right-hand diagram is

$$A' = G\left[\frac{1}{i} - \frac{n}{(1+i)^n - 1}\right]$$

In functional notation, the uniform gradient (to) uniform series factor is: $A' = G(A/G, i, n)$.

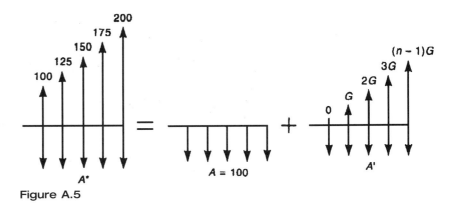

Figure A.5

The uniform gradient uniform series factor may be read from the compound interest tables directly, or computed as

$$(A/G, i, n) = \frac{1 - n(A/F, i, n)}{i}$$

Note carefully the diagrams for the uniform gradient factors. The first term in the uniform gradient is zero and the last term is $(n - 1)G$. But we use n in the equations and function notation. The derivations (not shown here) were done on this basis, and the uniform gradient compound interest tables are computed this way.

Example **A.11**

For the situation in Example A.10, we wish now to know the uniform annual maintenance cost. Compute an equivalent A for the maintenance costs.

Solution

Refer to Exhibit 2 The equivalent uniform annual maintenance cost is

$$A = 155 + 35(A/G, 6\%, 8) = 155 + 35(3.195) = \$266.83$$

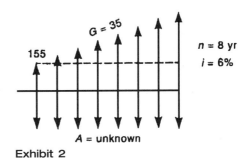

Exhibit 2

Standard compound interest tables give values for eight interest factors: two single payments, four uniform payment series, and two uniform gradients. The tables do *not* give the uniform gradient future worth factor, $(F/G, i, n)$. If it is needed, it may be computed from two tabulated factors:

$$(F/G, i, n) = (P/G, i, n)(F/P, i, n)$$

For example, if $i = 10$ percent and $n = 12$ years, then $(F/G, 10\%, 12) = (P/G, 10\%, 12)(F/P, 10\%, 12) = (29.901)(3.138) = 93.83$.

A second method of computing the uniform gradient future worth factor is

$$(F/G, i, n) = \frac{(F/A, i, n) - n}{i}$$

Using this equation for $i = 10$ percent and $n = 12$ years, $(F/G, 10\%, 12) = [(F/A, 10\%, 12) - 12]/0.10 = (21.384 - 12)/0.10 = 93.84.$

Continuous Compounding

Table A.2   Continuous compounding: Functional notation and formulas

Factor	Given	To Find	Functional Notation	Formula
Single payment				
Compound amount factor	P	F	$(F/P, r\%, n)$	$F = P[e^{rn}]$
Present worth factor	F	P	$(P/F, r\%, n)$	$P = F[e^{-rn}]$
Uniform payment series				
Sinking fund factor	F	A	$(A/F, r\%, n)$	$A = F\left[\frac{e^r - 1}{e^{rn} - 1}\right]$
Capital recovery factor	P	A	$(A/P, r\%, n)$	$A = P\left[\frac{e^r - 1}{1 - e^{-rn}}\right]$
Compound amount factor	A	F	$(F/A, r\%, n)$	$F = A\left[\frac{e^{rn} - 1}{e^r - 1}\right]$
Present worth factor	A	P	$(P/A, r\%, n)$	$P = A\left[\frac{1 - e^{-rn}}{e^r - 1}\right]$

r = nominal annual interest rate, n = number of years.

Example A.12

Five hundred dollars is deposited each year into a savings bank account that pays 5 percent nominal interest, compounded continuously. How much will be in the account at the end of five years?

Solution

$A = \$500$, $r = 0.05$, $n = 5$ years.

$$F = A(F/A, r\%, n) = A\left[\frac{e^{rn} - 1}{e^r - 1}\right] = 500\left[\frac{e^{0.05(5)} - 1}{e^{0.05} - 1}\right] = \$2769.84$$

NOMINAL AND EFFECTIVE INTEREST

Nominal interest is the annual interest rate without considering the effect of any compounding. *Effective interest* is the annual interest rate taking into account the effect of any compounding during the year.

Non-Annual Compounding

Frequently an interest rate is described as an annual rate, even though the interest period may be something other than one year. A bank may pay 1 percent interest on the amount in a savings account every three months. The *nominal* interest rate in this situation is $4 \times 1\% = 4\%$. But if you deposited $1000 in such an account, would you have $104\%(1000) = \$1040$ in the account at the end of one year? The answer is no, you would have more. The amount in the account would increase as follows:

Amount in Account

Beginning of year:	1000.00
End of three months:	$1000.00 + 1\%(1000.00) = 1010.00$
End of six months:	$1010.00 + 1\%(1010.00) = 1020.10$
End of nine months:	$1020.10 + 1\%(1020.10) = 1030.30$
End of one year:	$1030.30 + 1\%(1030.30) = 1040.60$

At the end of one year, the interest of $40.60, divided by the original $1000, gives a rate of 4.06 percent. This is the *effective* interest rate.

$$\text{Effective interest rate per year:} \quad i_{\text{eff}} = (1 + r/m)^m - 1$$

where r = nominal annual interest rate
m = number of compound periods per year
r/m = effective interest rate per period

Example A.13

A bank charges 1.5 percent interest per month on the unpaid balance for purchases made on its credit card. What nominal interest rate is it charging? What is the effective interest rate?

Solution

The nominal interest rate is simply the annual interest ignoring compounding, or $12(1.5\%) = 18\%$.

$$\text{Effective interest rate} = (1 + 0.015)^{12} - 1 = 0.1956 = 19.56\%$$

Continuous Compounding

When m, the number of compound periods per year, becomes very large and approaches infinity, the duration of the interest period decreases from Δt to dt. For this condition of *continuous compounding*, the effective interest rate per year is

$$i_{\text{eff}} = e^r - 1$$

where r = nominal annual interest rate.

Example A.14

If the bank in Example A.13 changes its policy and charges 1.5 percent per month, compounded continuously, what nominal and what effective interest rate is it charging?

Solution

Nominal annual interest rate, $r = 12 \times 1.5\% = 18\%$

Effective interest rate per year, $i_{\text{eff}} = e^{0.18} - 1 = 0.1972 = 19.72\%$

SOLVING ENGINEERING ECONOMICS PROBLEMS

The techniques presented so far illustrate how to convert single amounts of money, and uniform or gradient series of money, into some equivalent sum at another point in time. These compound interest computations are an essential part of engineering economics problems.

The typical situation is that we have a number of alternatives; the question is, which alternative should we select? The customary method of solution is to express each alternative in some common form and then choose the best, taking both the monetary and intangible factors into account. In most computations an interest rate must be used. It is often called the minimum attractive rate of return (MARR), to indicate that this is the smallest interest rate, or rate of return, at which one is willing to invest money.

Criteria

Engineering economics problems inevitably fall into one of three categories:

1. *Fixed input.* The amount of money or other input resources is fixed. *Example*: A project engineer has a budget of $450,000 to overhaul a plant.

2. *Fixed output.* There is a fixed task or other output to be accomplished. *Example*: A mechanical contractor has been awarded a fixed-price contract to air-condition a building.

3. *Neither input nor output fixed.* This is the general situation, where neither the amount of money (or other inputs) nor the amount of benefits (or other outputs) is fixed. *Example*: A consulting engineering firm has more work available than it can handle. It is considering paying the staff to work evenings to increase the amount of design work it can perform.

There are five major methods of comparing alternatives: present worth, future worth, annual cost, rate of return, and benefit-cost analysis. These are presented in the sections that follow.

PRESENT WORTH

Present worth analysis converts all of the money consequences of an alternative into an equivalent present sum. The criteria are

Category	Present Worth Criterion
Fixed input	Maximize the present worth of benefits or other outputs
Fixed output	Minimize the present worth of costs or other inputs
Neither input nor output fixed	Maximize present worth of benefits minus present worth of costs, or maximize net present worth

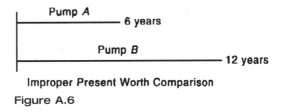

Improper Present Worth Comparison
Figure A.6

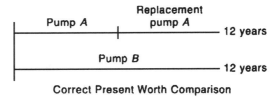

Correct Present Worth Comparison
Figure A.7

Appropriate Problems

Present worth analysis is most frequently used to determine the present value of future money receipts and disbursements. We might want to know, for example, the present worth of an income-producing property, such as an oil well. This should provide an estimate of the price at which the property could be bought or sold.

An important restriction in the use of present worth calculation is that there must be a common analysis period for comparing alternatives. It would be incorrect, for example, to compare the present worth (PW) of cost of pump A, expected to last 6 years, with the PW of cost of pump B, expected to last 12 years (Figure A.6). In situations like this, the solution is either to use some other analysis technique (generally, the annual cost method is suitable in these situations) or to restructure the problem so that there is a common analysis period.

In this example, a customary assumption would be that a pump is needed for 12 years and that pump A will be replaced by an identical pump A at the end of 6 years. This gives a 12-year common analysis period (Figure A.7). This approach is easy to use when the different lives of the alternatives have a practical least-common-multiple life. When this is not true (for example, the life of J equals 7 years and the life of K equals 11 years), some assumptions must be made to select a suitable common analysis period, or the present worth method should not be used.

Example **A.15**

Machine X has an initial cost of $10,000, an annual maintenance cost of $500 per year, and no salvage value at the end of its 4-year useful life. Machine Y costs $20,000, and the first year there is no maintenance cost. Maintenance is $100 the second year, and it increases $100 per year thereafter. The machine has an anticipated $5000 salvage value at the end of its 12-year useful life. If the minimum attractive rate of return (MARR) is 8 percent, which machine should be selected?

Solution

The analysis period is not stated in the problem. Therefore, we select the least common multiple of the lives, or 12 years, as the analysis period.

Present worth of cost of 12 years of machine X:

$$PW = 10{,}000 + 10{,}000(P/F, 8\%, 4) + 10{,}000(P/F, 8\%, 8) + 500(P/A, 8\%, 12)$$
$$= 10{,}000 + 10{,}000(0.7350) + 10{,}000(0.5403) + 500(7.536) = \$26{,}521$$

Present worth of cost of 12 years of machine Y:

$$PW = 20{,}000 + 100(P/G, 8\%, 12) - 5000(P/F, 8\%, 12)$$
$$= 20{,}000 + 100(34.634) - 5000(0.3971) = \$21{,}478$$

Choose machine Y, with its smaller PW of cost.

Example A.16

Two alternatives have the following cash flows:

	Alternative	
Year	A	B
0	−$2000	−$2800
1	+800	+1100
2	+800	+1100
3	+800	+1100

At a 4 percent interest rate, which alternative should be selected?

Solution

The net present worth of each alternative is computed:

Net present worth (NPW) = PW of benefit − PW of cost
$$NPW_A = 800(P/A, 4\%, 3) - 2000 = 800(2.775) - 2000 = \$220.00$$
$$NPW_B = 1100(P/A, 4\%, 3) - 2800 = 1100(2.775) - 2800 = \$252.50$$

To maximize NPW, choose alternative B.

Infinite Life and Capitalized Cost

In the special situation where the analysis period is infinite ($n = \infty$), an analysis of the present worth of cost is called *capitalized cost*. There are a few public projects where the analysis period is infinity. Other examples are permanent endowments and cemetery perpetual care.

When n equals infinity, a present sum P will accrue interest of Pi for every future interest period. For the principal sum P to continue undiminished (an essential requirement for n equal to infinity), the end-of-period sum A that can be disbursed is Pi (Figure A.8). When $n = \infty$, the fundamental relationship is

$$A = Pi$$

Some form of this equation is used whenever there is a problem involving an infinite analysis period.

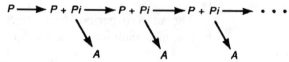

Figure A.8

In his will, a man wishes to establish a perpetual trust to provide for the maintenance of a small local park. If the annual maintenance is $7500 per year and the trust account can earn 5 percent interest, how much money must be set aside in the trust?

Solution

When $n = \infty$, $A = Pi$ or $P = A/i$. The capitalized cost is $P = A/i = \$7500/0.05 = \$150,000$.

FUTURE WORTH OR VALUE

In present worth analysis, the comparison is made in terms of the equivalent present costs and benefits. But the analysis need not be made in terms of the present—it can be made in terms of a past, present, or future time. Although the numerical calculations may look different, the decision is unaffected by the selected point in time. Often we do want to know what the future situation will be if we take some particular couse of action now. An analysis based on some future point in time is called *future worth analysis.*

Category	Future Worth Criterion
Fixed input	Maximize the future worth of benefits or other outputs
Fixed output	Minimize the future worth of costs or other inputs
Neither input nor output fixed	Maximize future worth of benefits minus future worth of costs, or maximize net future worth

Example **A.18**

Two alternatives have the following cash flows:

	Alternative	
Year	A	B
0	−$2000	−$2800
1	+800	+1100
2	+800	+1100
3	+800	+1100

At a 4 percent interest rate, which alternative should be selected?

Solution

In Example A.16, this problem was solved by present worth analysis at year 0. Here it will be solved by future worth analysis at the end of year 3.

Net future worth (NFW) = FW of benefits − FW of cost

$$\text{NFW}_A = 800(F/A, 4\%, 3) - 2000(F/P, 4\%, 3)$$
$$= 800(3.122) - 2000(1.125) = +\$247.60$$

$$\text{NFW}_B = 1100(F/A, 4\%, 3) - 2800(F/P, 4\%, 3)$$
$$= 1100(3.122) - 2800(1.125) = +\$284.20$$

To maximize NFW, choose alternative *B*.

ANNUAL COST

The annual cost method is more accurately described as the method of equivalent uniform annual cost (EUAC). Where the computation is of benefits, it is called the method of equivalent uniform annual benefits (EUAB).

Criteria

For each of the three possible categories of problems, there is an annual cost criterion for economic efficiency.

Category	Annual Cost Criterion
Fixed input	Maximize the equivalent uniform annual benefits (EUAB)
Fixed output	Minimize the equivalent uniform annual cost (EUAC)
Neither input nor output fixed	Maximize EUAB − EUAC

Application of Annual Cost Analysis

In the section on present worth, we pointed out that the present worth method requires a common analysis period for all alternatives. This restriction does not apply in all annual cost calculations, but it is important to understand the circumstances that justify comparing alternatives with different service lives.

Frequently, an analysis is done to provide for a more-or-less continuing requirement. For example, one might need to pump water from a well on a continuing basis. Regardless of whether each of two pumps has a useful service life of 6 years or 12 years, we would select the alternative whose annual cost is a minimum. And this still would be the case if the pumps' useful lives were the more troublesome 7 and 11 years. Thus, if we can assume a continuing need for an item, an annual cost comparison among alternatives of differing service lives is valid. The underlying assumption in these situations is that the shorter-lived alternative can be replaced with an identical item with identical costs, when it has reached the end of its useful life. This means that the EUAC of the initial alternative is equal to the EUAC for the continuing series of replacements.

On the other hand, if there is a specific requirement to pump water for 10 years, then each pump must be evaluated to see what costs will be incurred during the analysis period and what salvage value, if any, may be recovered at the end of the analysis period. The annual cost comparison needs to consider the actual circumstances of the situation.

Examination problems are often readily solved using the annual cost method. And the underlying "continuing requirement" is usually present, so an annual cost comparison of unequal-lived alternatives is an appropriate method of analysis.

Example **A.19**

Consider the following alternatives:

	A	B
First cost	$5000	$10,000
Annual maintenance	500	200
End-of-useful-life salvage value	600	1000
Useful life	5 years	15 years

Based on an 8 percent interest rate, which alternative should be selected?

Solution

Assuming both alternatives perform the same task and there is a continuing requirement, the goal is to minimize EUAC.

Alternative *A:*

$$EUAC = 5000(A/P, 8\%, 5) + 500 - 600(A/F, 8\%, 5)$$
$$= 5000(0.2505) + 500 - 600(0.1705) = \$1650$$

Alternative *B:*

$$EUAC = 10,000(A/P, 8\%, 15) + 200 - 1000(A/F, 8\%, 15)$$
$$= 10,000(0.1168) + 200 - 1000(0.0368) = \$1331$$

To minimize EUAC, select alternative *B*.

RATE OF RETURN ANALYSIS

A typical situation is a cash flow representing the costs and benefits. The rate of return may be defined as the interest rate where PW of cost = PW of benefits, EUAC = EUAB, or PW of cost – PW of benefits = 0.

Example **A.20**

Compute the rate of return for the investment represented by the following cash flow table.

Year:	0	1	2	3	4	5
Cash flow:	–$595	+250	+200	+150	+100	+50

Solution

This declining uniform gradient series may be separated into two cash flows (Exhibit 3) for which compound interest factors are available.

Note that the gradient series factors are based on an *increasing* gradient. Here the declining cash flow is solved by subtracting an increasing uniform gradient, as indicated in the figure.

PW of cost – PW of benefits = 0

$$595 - [250(P/A, i, 5) - 50(P/G, i, 5) = 0$$

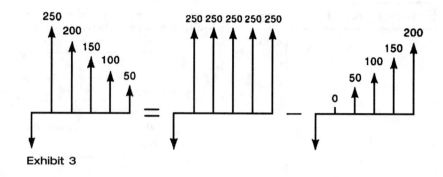

Exhibit 3

Try $i = 10\%$:

$$595 - [250(3.791) - 50(6.862)] = -9.65$$

Try $i = 12\%$:

$$595 - [250(3.605) - 50(6.397)] = +13.60$$

The rate of return is between 10 percent and 12 percent. It may be computed more accurately by linear interpolation:

$$\text{Rate of return} = 10\% + (2\%)\left(\frac{9.65 - 0}{13.60 + 9.65}\right) = 10.83\%.$$

Two Alternatives

Compute the incremental rate of return on the cash flow representing the difference between the two alternatives. Since we want to look at increments of *investment*, the cash flow for the difference between the alternatives is computed by taking the higher initial-cost alternative minus the lower initial-cost alternative. If the incremental rate of return is greater than or equal to the predetermined minimum attractive rate of return (MARR), choose the higher-cost alternative; otherwise, choose the lower-cost alternative.

Example A.21

Two alternatives have the following cash flows:

| | Alternative | |
Year	A	B
0	-$2000	-$2800
1	+800	+1100
2	+800	+1100
3	+800	+1100

If 4 percent is considered the minimum attractive rate of return (MARR), which alternative should be selected?

Solution

These two alternatives were previously examined in Examples A.16 and A.18 by present worth and future worth analysis. This time, the alternatives will be resolved using a rate-of-return analysis.

Note that the problem statement specifies a 4 percent MARR, whereas Examples A.16 and A.18 referred to a 4 percent interest rate. These are really two different ways of saying the same thing: The minimum acceptable time value of money is 4 percent.

First, tabulate the cash flow that represents the increment of investment between the alternatives. This is done by taking the higher initial-cost alternative minus the lower initial-cost alternative:

	Alternative		Difference Between Alternatives
Year	A	B	B – A
0	–$2000	–$2800	–$800
1	+800	+1100	+300
2	+800	+1100	+300
3	+800	+1100	+300

Then compute the rate of return on the increment of investment represented by the difference between the alternatives:

$$\text{PW of cost} = \text{PW of benefits}$$
$$800 = 300(P/A, i, 3)$$

$$(P/A, i, 3) = 800/300 = 2.67$$
$$i = 6.1\%$$

Since the incremental rate of return exceeds the 4 percent MARR, the increment of investment is desirable. Choose the higher-cost alternative *B*.

Before leaving this example, one should note something that relates to the rates of return on alternative *A* and on alternative *B*. These rates of return, if calculated, are

	Rate of Return
Alternative *A*	9.7%
Alternative *B*	8.7%

The correct answer to this problem has been shown to be alternative *B*, even though alternative *A* has a higher rate of return. The higher-cost alternative may be thought of as the lower-cost alternative plus the increment of investment between them. Viewed this way, the higher-cost alternative *B* is equal to the desirable lower-cost alternative *A* plus the difference between the alternatives.

The important conclusion is that computing the rate of return for each alternative does *not* provide the basis for choosing between alternatives. Instead, incremental analysis is required.

Example **A.22**

Consider the following:

	Alternative	
Year	A	B
0	−$200.0	−$131.0
1	+77.6	+48.1
2	+77.6	+48.1
3	+77.6	+48.1

If the MARR is 10 percent, which alternative should be selected?

Solution

To examine the increment of investment between the alternatives, we will examine the higher initial-cost alternative minus the lower initial-cost alternative, or $A - B$.

	Alternative		Increment
Year	A	B	A − B
0	−$200.0	−$131.0	−$69.0
1	+77.6	+48.1	+29.5
2	+77.6	+48.1	+29.5
3	+77.6	+48.1	+29.5

Solve for the incremental rate of return:

$$\text{PW of cost} = \text{PW of benefits}$$
$$69.0 = 29.5(P/A, i, 3)$$
$$(P/A, i, 3) = 69.0/29.5 = 2.339$$

From compound interest tables, the incremental rate of return is between 12 percent and 18 percent. This is a desirable increment of investment; hence we select the higher-initial-cost alternative A.

Three or More Alternatives

When there are three or more mutually exclusive alternatives, proceed with the same logic presented for two alternatives. The components of incremental analysis are listed below.

Step 1. Compute the rate of return for each alternative. Reject any alternative where the rate of return is less than the desired MARR. (This step is not essential, but helps to immediately identify unacceptable alternatives.)

Step 2. Rank the remaining alternatives in order of increasing initial cost.

Step 3. Examine the increment of investment between the two lowest-cost alternatives as described for the two-alternative problem. Select the better of the two alternatives and reject the other one.

Step 4. Take the preferred alternative from step 3. Consider the next higher initial-cost alternative and proceed with another two-alternative comparison.

Step 5. Continue until all alternatives have been examined and the best of the multiple alternatives has been identified.

Example **A.23**

Consider the following:

	Alternative	
Year	A	B
0	−$200.0	−$131.0
1	+77.6	+48.1
2	+77.6	+48.1
3	+77.6	+48.1

If the MARR is 10 percent, which alternative, if any, should be selected?

Solution

One should carefully note that this is a *three-alternative* problem, where the alternatives are A, B, and *Do nothing*. In this solution we will skip step 1. Reorganize the problem by placing the alternatives in order of increasing initial cost:

	Alternative		
Year	Do Nothing	B	A
0	0	−$131.0	−$200.0
1	0	+48.1	+77.6
2	0	+48.1	+77.6
3	0	+48.1	+77.6

Examine the *B − Do nothing* increment of investment:

Year	B − Do Nothing
0	−$131.0 − 0 = −$131.0
1	+48.1 − 0 = +48.1
2	+48.1 − 0 = +48.1
3	+48.1 − 0 = +48.1

Solve for the incremental rate of return:

$$\text{PW of cost} = \text{PW of benefits}$$
$$131.0 = 48.1(P/A, \, i, \, 3)$$
$$(P/A, \, i, \, 3) = 131.0/48.1 = 2.723$$

From compound interest tables, the incremental rate of return is about 5 percent. Since the incremental rate of return is less than 10 percent, the *B − Do nothing* increment is not desirable. Reject alternative B.

Year	A − Do Nothing
0	−$200.0 − 0 = −$200.0
1	+77.6 − 0 = +77.6
2	+77.6 − 0 = +77.6
3	+77.6 − 0 = +77.6

Next, consider the increment of investment between the two remaining alternatives. Solve for the incremental rate of return:

$$PW \text{ of cost} = PW \text{ of benefits}$$
$$200.0 = 77.6(P/A, i, 3)$$
$$(P/A, i, 3) = 200.0/77.6 = 2.577$$

The incremental rate of return is 8 percent, less than the desired 10 percent. Reject the increment and select the remaining alternative: *Do nothing*.

If you have not already done so, you should go back to Example A.22 and see how the slightly changed wording of the problem has radically altered it. Example A.22 required a choice between two undesirable alternatives. This example adds the *Do nothing* alternative, which is superior to A and B.

BENEFIT-COST ANALYSIS

Generally, in public works and governmental economic analyses, the dominant method of analysis is the *benefit-cost ratio*. It is simply the ratio of benefits divided by costs, taking into account the time value of money.

$$B/C = \frac{PW \text{ of benefits}}{PW \text{ of cost}} = \frac{\text{Equivalent uniform annual benefits}}{\text{Equivalent uniform annual cost}}$$

For a given interest rate, a B/C ratio ≥ 1 reflects an acceptable project. The B/C analysis method is parallel to rate-of-return analysis. The same kind of incremental analysis is required.

Example A.24

Solve Example A.22 by benefit-cost analysis.

Solution

	Alternative		Increment
Year	A	B	A − B
0	− $200.0	− $131.0	− $69.0
1	+ 77.6	+ 48.1	+ 29.5
2	+ 77.6	+ 48.1	+ 29.5
3	+ 77.6	+ 48.1	+ 29.5

The benefit-cost ratio for the A − B increment is

$$B/C = \frac{PW \text{ of benefits}}{PW \text{ of cost}} = \frac{29.5(P/A, 10\%, 3)}{69.0} = \frac{73.37}{69.0} = 1.06$$

Since the B/C ratio exceeds 1, the increment of investment is desirable. Select the higher-cost alternative A.

BREAKEVEN ANALYSIS

In business, "breakeven" is defined as the point where income just covers costs. In engineering economics, the breakeven point is defined as the point where two alternatives are equivalent.

Example A.25

A city is considering a new $50,000 snowplow. The new machine will operate at a savings of $600 per day compared with the present equipment. Assume that the MARR is 12 percent, and the machine's life is 10 years with zero resale value at that time. How many days per year must the machine be used to justify the investment?

Solution

This breakeven problem may be readily solved by annual cost computations. We will set the equivalent uniform annual cost (EUAC) of the snowplow equal to its annual benefit and solve for the required annual utilization. Let X = breakeven point = days of operation per year.

$$\text{EUAC} = \text{EUAB}$$
$$50,000(A/P,\ 12\%,\ 10) = 600X$$
$$X = 50,000(0.1770)/600 = 14.8 \text{ days/year}$$

OPTIMIZATION

Optimization is the determination of the best or most favorable situation.

Minima-Maxima

In problems where the situation can be represented by a function, the customary approach is to set the first derivative of the function to zero and solve for the root(s) of this equation. If the second derivative is *positive*, the function is a minimum for the critical value; if it is *negative*, the function is a maximum.

Example A.26

A consulting engineering firm estimates that their net profit is given by the equation

$$P(x) = -0.03x^3 + 36x + 500 \quad x \geq 0$$

where x = number of employees and $P(x)$ = net profit. What is the optimal number of employees?

Solution

$$P'(x) = -0.09x^2 + 36 = 0 \quad P''(x) = -0.18x$$
$$x^2 = 36/0.09 = 400$$
$$x = 20 \text{ employees.}$$
$$P''(20) = -0.18(20) = -3.6$$

Since $P''(20) < 0$, the net profit is maximized for 20 employees.

Economic Problem—Best Alternative

Since engineering economics problems seek to identify the best or most favorable situation, they are by definition optimization problems. Most use compound interest computations in their solution, but some do not. Consider the following example.

Example A.27

A firm must decide which of three alternatives to adopt to expand its capacity. It wants a minimum annual profit of 20 percent of the initial cost of each increment of investment. Any money not invested in capacity expansion can be invested elsewhere for an annual yield of 20 percent of the initial cost.

Alternative	Initial Cost	Annual Profit	Profit Rate
A	$100,000	$30,000	30%
B	300,000	66,00	22
C	500,000	80,000	16

Which alternative should be selected?

Solution

Since alternative C fails to produce the 20 percent minimum annual profit, it is rejected. To decide between alternatives A and B, examine the profit rate for the B − A increment.

Alternative	Initial Cost	Annual Profit	Incremental Cost	Incremental Profit	Incremental Profit Rate
A	$100,000	$30,000			
			$200,000	$36,000	18%
B	300,000	66,000			

The B − A incremental profit rate is less than the minimum 20 percent, so alternative B should be rejected. Thus the best investment of $300,000, for example, would be alternative A (annual profit = $30,000) plus $200,000 invested elsewhere at 20 percent (annual profit = $40,000). This combination would yield a $70,000 annual profit, which is better than the alternative B profit of $66,000. Select A.

Economic Order Quantity

One special case of optimization occurs when an item is used continuously and is periodically purchased. Thus the inventory of the item fluctuates from zero (just prior to the receipt of the purchased quantity) to the purchased quantity (just after receipt). The simplest model for the economic order quantity (EOQ) is

$$EOQ = \sqrt{\frac{2BD}{E}}$$

where

B = ordering cost, \$/order
D = demand per period, units
E = inventory holding cost, \$/unit/period
EOC = economic order quantity, units

Example A.28

A company uses 8000 wheels per year in its manufacture of golf carts. The wheels cost \$15 each and are purchased from an outside supplier. The money invested in the inventory costs 10 percent per year, and the warehousing cost amounts to an additional 2 percent per year. It costs \$150 to process each purchase order. When an order is placed, how many wheels should be ordered?

Solution

$$\text{EOQ} = \sqrt{\frac{2 \times \$150 \times 8000}{(10\% + 2\%)(15.00)}} = 1155 \text{ wheels}$$

VALUATION AND DEPRECIATION

Depreciation of capital equipment is an important component of many after-tax economic analyses. For this reason, one must understand the fundamentals of depreciation accounting.

Notation

BV = book value
C = cost of the property (basis)
D_j = depreciation in year j
S_n = salvage value in year n

Depreciation is the systematic allocation of the cost of a capital asset over its useful life. *Book value* is the original cost of an asset, minus the accumulated depreciation of the asset.

$$\text{BV} = C - \Sigma(D_j)$$

In computing a schedule of depreciation charges, four items are considered.

1. Cost of the property, C (called the *basis* in tax law).

2. Type of property. Property is classified as either *tangible* (such as machinery) or *intangible* (such as a franchise or a copyright), and as either *real property* (real estate) or *personal property* (everything that is not real property).

3. Depreciable life in years, n.

4. Salvage value of the property at the end of its depreciable (useful) life, S_n.

Straight-Line Depreciation

The depreciation charge in any year is

$$D_j = \frac{C - S_n}{n}$$

An alternative computation is

$$\text{Depreciation charge in any year, } D_j = \frac{C - \text{depreciation taken to beginning of year } j - S_n}{\text{Remaining useful life at beginning of year } j}$$

Sum-of-Years'-Digits Depreciation

$$\text{Depreciation charge in any year, } D_j = \frac{\text{Remaining depreciable life at beginning of year}}{\text{Sum of years' digits for total useful life}} \times (C - S_n)$$

Declining-Balance Depreciation

$$\text{Double declining-balance depreciation charge in any year, } D_j = \frac{2C}{m}\left(1 - \frac{2}{n}\right)^{j-1}$$

$$\text{Total depreciation at the end of } n \text{ years, } C = \left[1 - \left(1 - \frac{2}{n}\right)^n\right]$$

$$\text{Book value at the end of } j \text{ years, } BV_j = C\left(1 - \frac{2}{n}\right)^j$$

For 150 percent declining-balance depreciation, replace the 2 in the three equations above with 1.5.

Sinking-Fund Depreciation

$$\text{Depreciation charge in any year, } D_j = (C - S_n)(A/F, i\%, n)(F/P, i\%, j - 1)$$

Modified Accelerated Cost Recovery System Depreciation

The modified accelerated cost recovery system (MACRS) depreciation method generally applies to property placed in service after 1986. To compute the MACRS depreciation for an item, one must know

1. Cost (basis) of the item.

2. Property class. All tangible property is classified in one of six classes (3, 5, 7, 10, 15, and 20 years), which is the life over which it is depreciated (see Table A.3). Residential real estate and nonresidential real estate are in two separate real property classes of 27.5 years and 39 years, respectively.

3. Depreciation computation.

Table A.3 MACRS classes of depreciable property

Property Class	Personal Property (All Property Except Real Estate)
3-year property	Special handling devices for food and beverage manufacture Special tools for the manufacture of finished plastic products, fabricated metal products, and motor vehicles Property with an asset depreciation range (ADR) midpoint life of 4 years or less
5-year property	Automobiles* and trucks Aircraft (of non–air-transport companies) Equipment used in research and experimentation Computers Petroleum drilling equipment Property with an ADR midpoint life of more than 4 years and less than 10 years
7-year property	All other property not assigned to another class Office furniture, fixtures, and equipment Property with an ADR midpoint life of 10 years or more, and less than 16 years
10-year property	Assets used in petroleum refining and preparation of certain food products Vessels and water transportation equipment Property with an ADR midpoint life of 16 years or more, and less than 20 years
15-year property	Telephone distribution plants Municipal sewage treatment plants Property with an ADR midpoint life of 20 years or more, and less than 25 years
20-year property	Municipal sewers Property with an ADR midpoint life of 25 years or more

Property Class	Real Property (Real Estate)
27.5 years	Residential rental property (does not include hotels and motels)
39 years	Nonresidential real property

*The depreciation deduction for automobiles is limited to $2860 in the first tax year and is reduced in subsequent years.

- Use double-declining-balance depreciation for 3-, 5-, 7-, and 10-year property classes with conversion to straight-line depreciation in the year that increases the deduction.

- Use 150%-declining-balance depreciation for 15- and 20-year property classes with conversion to straight-line depreciation in the year that increases the deduction.

- In MACRS, the salvage value is assumed to be zero.

Half-Year Convention

Except for real property, a half-year convention is used. Under this convention all property is considered to be placed in service in the middle of the tax year, and a half-year of depreciation is allowed in the first year. For each of the remaining years, one is allowed a full year of depreciation. If the property is disposed of

Table A.4 MACRS* depreciation for personal property—half-year convention

If the Recovery Year Is	The Applicable Percentage for the Class of Property Is			
	3-Year Class	5-Year Class	7-Year Class	10-Year Class
1	33.33	20.00	14.29	10.00
2	44.45	32.00	24.49	18.00
3	14.81†	19.20	17.49	14.40
4	7.41	11.52†	12.49	11.52
5		11.52	8.93†	9.22
6		5.76	8.92	7.37
7			8.93	6.55†
8			4.46	6.55
9				6.56
10				6.55
11				3.28

*In the *Fundamentals of Engineering Reference Handbook*, this table is called "Modified ACRS Factors."

†Use straight-line depreciation for the year marked and all subsequent years.

prior to the end of the recovery period (property class life), a half-year of depreciation is allowed in that year. If the property is held for the entire recovery period, a half-year of depreciation is allowed for the year following the end of the recovery period (see Table A.4). Owing to the half-year convention, a general form of the double-declining-balance computation must be used to compute the year-by-year depreciation.

DDB depreciation in any year, $D_j = \dfrac{2}{n}(C - \text{depreciation in years prior to } j)$

Example A.29

A \$5000 computer has an anticipated \$500 salvage value at the end of its five-year depreciable life. Compute the depreciation schedule for the machinery by (a) sum-of-years'-digits depreciation and (b) MACRS depreciation. Do the MACRS computation by hand, and then compare the results with the values from Table A.4.

Solution

(a) Sum-of-years'-digits depreciation:

$$D_j = \frac{n-j+1}{\frac{n}{2}(n+1)}(C - S_n)$$

$$D_1 = \frac{5-1+1}{\frac{5}{2}(5+1)}(5000 - 500) = \$1500$$

$$D_2 = \frac{5-2+1}{\frac{5}{2}(5+1)}(5000-500) = \$1200$$

$$D_3 = \frac{5-3+1}{\frac{5}{2}(5+1)}(5000-500) = 900$$

$$D_4 = \frac{5-4+1}{\frac{5}{2}(5+1)}(5000-500) = 600$$

$$D_5 = \frac{5-5+1}{\frac{5}{2}(5+1)}(5000-500) = 300$$

$$\overline{\$4500}$$

(b) MACRS depreciation. Double-declining-balance with conversion to straight-line. Five-year property class. Half-year convention. Salvage value S_n is assumed to be zero for MACRS. Using the general DDB computation,

Year

1 (half-year) $\quad D_1 = \frac{1}{2} \times \frac{2}{5}(5000-0) \quad = \$1000$

2 $\qquad D_2 = \frac{2}{5}(5000-1000) \quad = 1600$

3 $\qquad D_3 = \frac{2}{5}(5000-2600) \quad = 960$

4 $\qquad D_4 = \frac{2}{5}(5000-3560) \quad = 576$

5 $\qquad D_5 = \frac{2}{5}(5000-4136) \quad = 346$

6 (half-year) $\quad D_6 = \frac{1}{2} \times \frac{2}{5}(5000-4482) = 104$

$$\overline{\$4586}$$

The computation must now be modified to convert to straight-line depreciation at the point where the straight-line depreciation will be larger. Using the alternative straight-line computation,

$$D_5 = \frac{5000-4136-0}{1.5 \text{ years remaining}} = \$576$$

This is more than the $346 computed using DDB, hence switch to straight-line for year 5 and beyond.

$$D_6 \text{ (half-year)} = \frac{1}{2}(576) = \$288$$

Answers:

	Depreciation	
Year	SOYD	MACRS
1	$1500	$1000
2	1200	1600
3	900	960
4	600	576
5	300	576
6	0	288
	$4500	$5000

The computed MACRS depreciation is identical to the result obtained from Table A.4.

TAX CONSEQUENCES

Income taxes represent another of the various kinds of disbursements encountered in an economic analysis. The starting point in an after-tax computation is the before-tax cash flow. Generally, the before-tax cash flow contains three types of entries:

1. Disbursements of money to purchase capital assets. These expenditures create no direct tax consequence, for they are the exchange of one asset (money) for another (capital equipment).

2. Periodic receipts and/or disbursements representing operating income and/or expenses. These increase or decrease the year-by-year tax liability of the firm.

3. Receipts of money from the sale of capital assets, usually in the form of a salvage value when the equipment is removed. The tax consequences depend on the relationship between the book value (cost – depreciation taken) of the asset and its salvage value.

Situation	Tax Consequence
Salvage value > Book value	Capital gain on differences
Salvage value = Book value	No tax consequence
Salvage value < Book value	Capital loss on difference

After determining the before-tax cash flow, compute the depreciation schedule for any capital assets. Next, compute taxable income, the taxable component of the before-tax cash flow minus the depreciation. The income tax is the taxable income times the appropriate tax rate. Finally, the after-tax cash flow is the before-tax cash flow adjusted for income taxes.

To organize these data, it is customary to arrange them in the form of a cash flow table, as follows:

Year	Before-Tax Cash Flow	Depreciation	Taxable Income	Income Taxes	After-Tax Cash Flow
0	•				•
1	•	•	•	•	•

Example **A.30**

A corporation expects to receive $32,000 each year for 15 years from the sale of a product. There will be an initial investment of $150,000. Manufacturing and sales expenses will be $8067 per year. Assume straight-line depreciation, a 15-year useful life, and no salvage value. Use a 46 percent income tax rate. Determine the projected after-tax rate of return.

Solution

Straight-line depreciation, $D_j = \dfrac{C - S_n}{n} = \dfrac{\$150,000 - 0}{15} = \$10,000$ per year

Year	Before-Tax Cash Flow	Depreciation	Taxable Income	Income Taxes	After-Tax Cash Flow
0	−150,000				−150,000
1	+23,933	10,000	13,933	−6,409	+17,524
2	+23,933	10,000	13,933	−6,409	+17,524
•	•	•	•	•	•
•	•	•	•	•	•
•	•	•	•	•	•
15	+23,933	10,000	13,933	−6,409	+17,524

Take the after-tax cash flow and compute the rate of return at which the PW of cost equals the PW of benefits.

$$150,000 = 17,524(P/A, i\%, 15)$$

$$(P/A, i\%, 15) = \frac{150,000}{17,524} = 8.559$$

From the compound interest tables, the after-tax rate of return is $i = 8\%$.

INFLATION

Inflation is characterized by rising prices for goods and services, whereas deflation produces a fall in prices. An inflationary trend makes future dollars have less purchasing power than present dollars. This helps long-term borrowers of money, for they may repay a loan of present dollars in the future with dollars of reduced buying power. The help to borrowers is at the expense of lenders. Deflation has the opposite effect. Money borrowed at one point in time, followed by a deflationary period, subjects the borrower to loan repayment with dollars of greater purchasing power than those borrowed. This is to the lenders' advantage at the expense of borrowers.

Price changes occur in a variety of ways. One method of stating a price change is as a uniform rate of price change per year.

f = General inflation rate per interest period
i = Effective interest rate per interest period

The following situation will illustrate the computations. A mortgage is to be repaid in three equal payments of $5000 at the end of years 1, 2, and 3. If the annual inflation rate, f, is 8% during this period, and a 12% annual interest rate (*i*) is desired, what is the maximum amount the investor would be willing to pay for the mortgage?

The computation is a two-step process. First, the three future payments must be converted to dollars with the same purchasing power as today's (year 0) dollars.

Year	Actual Cash Flow		Multiplied by		Cash Flow Adjusted to Today's (yr. 0) Dollars
0	—		—		—
1	+5000	×	$(1+0.08)^{-1}$	=	+4630
2	+5000	×	$(1+0.08)^{-2}$	=	+4286
3	+5000	×	$(1+0.08)^{-3}$	=	+3969

The general form of the adjusting multiplier is

$$(1+f)^{-n} = (P/F, f, n)$$

Now that the problem has been converted to dollars of the same purchasing power (today's dollars, in this example), we can proceed to compute the present worth of the future payments.

Year	Adjusted Cash Flow		Multiplied by		Present Worth
0	—		—		—
1	+4630	×	$(1+0.12)^{-1}$	=	+4134
2	+4286	×	$(1+0.12)^{-2}$	=	+3417
3	+3969	×	$(1+0.12)^{-3}$	=	+2825
					$10,376

The general form of the discounting multiplier is

$$(1+i)^{-n} = (P/F, i\%, n)$$

Alternative Solution

Instead of doing the inflation and interest rate computations separately, one can compute a combined equivalent interest rate, *d*.

$$d = (1+f)(1+i) - 1 = i + f + i(f)$$

For this cash flow, $d = 0.12 + 0.08 + 0.12(0.08) = 0.2096$. Since we do not have 20.96 percent interest tables, the problem has to be calculated using present worth equations.

$$PW = 5000(1 + 0.2096)^{-1} + 5000(1 + 0.2096)^{-2} + 5000(1 + 0.2096)^{-3}$$
$$= 4134 + 3417 + 2825 = \$10,376$$

Example A.31

One economist has predicted that there will be 7 percent per year inflation of prices during the next 10 years. If this proves to be correct, an item that presently sells for $10 would sell for what price 10 years hence?

Solution

$$f = 7\%, \ P = \$10$$
$$F = ?, \ n = 10 \text{ years}$$

Here the computation is to find the future worth F, rather than the present worth, P.

$$F = P(1 + f)^{10} = 10(1 + 0.07)^{10} = \$19.67$$

Effect of Inflation on Rate of Return

The effect of inflation on the computed rate of return for an investment depends on how future benefits respond to the inflation. If benefits produce constant dollars, which are not increased by inflation, the effect of inflation is to reduce the before-tax rate of return on the investment. If, on the other hand, the dollar benefits increase to keep up with the inflation, the before-tax rate of return will not be adversely affected by the inflation.

This is not true when an after-tax analysis is made. Even if the future benefits increase to match the inflation rate, the allowable depreciation schedule does not increase. The result will be increased taxable income and income tax payments. This reduces the available after-tax benefits and, therefore, the after-tax rate of return.

Example **A.32**

A man bought a 5 percent tax-free municipal bond. It cost $1000 and will pay $50 interest each year for 20 years. The bond will mature at the end of 20 years and return the original $1000. If there is 2% annual inflation during this period, what rate of return will the investor receive after considering the effect of inflation?

Solution

$$d = 0.05, \ i = \text{unknown}, \ j = 0.02$$
$$d = i + j + i(j)$$
$$0.05 = i + 0.02 + 0.02i$$
$$1.02i = 0.03, \ i = 0.294 = 2.94\%$$

RISK ANALYSIS

Probability

Probability can be considered to be the long-run relative frequency of occurrence of an outcome. There are two possible outcomes from flipping a coin (a head or a tail). If, for example, a coin is flipped over and over, we can expect in the long run that half the time heads will appear and half the time tails will appear. We would say the probability of flipping a head is 0.50 and of flipping a tail is 0.50. Since the probabilities are defined so that the sum of probabilities for all possible outcomes is 1, the situation is

$$\text{Probability of flipping a head} = 0.50$$
$$\text{Probability of flipping a tail} = 0.50$$
$$\text{Sum of all possible outcomes} = \overline{1.00}$$

Example A.33

If one were to roll one die (that is, one-half of a pair of dice), what is the probability that either a 1 or a 6 would result?

Solution

Since a die is a perfect six-sided cube, the probability of any side appearing is 1/6.

$$\text{Probability of rolling a } 1 = P(1) = 1/6$$
$$2 = P(2) = 1/6$$
$$3 = P(3) = 1/6$$
$$4 = P(4) = 1/6$$
$$5 = P(5) = 1/6$$
$$6 = P(6) = 1/6$$

Sum of all possible outcomes = 6/6 = 1. The probability of rolling a 1 or a 6 = 1/6 + 1/6 = 1/3.

   In the preceding examples, the probability of each outcome was the same. This need not be the case.

Example A.34

In the game of blackjack, a perfect hand is a 10 or a face card plus an ace. What is the probability of being dealt a 10 or a face card from a newly shuffled deck of 52 cards? What is the probability of being dealt an ace in this same situation?

Solution

The three outcomes examined are to be dealt a 10 or a face card, an ace, or some other card. Every card in the deck represents one of these three possible outcomes. There are 4 aces; 16 10s, jacks, queens, and kings; and 32 other cards.

$$\text{Probability of being dealt a 10 or a face card} = 16/52 = 0.31$$
$$\text{Probability of being dealt an ace} = 4/52 = 0.08$$
$$\underline{\text{Probability of being dealt some other card} = 32/52 = 0.61}$$
$$1.00$$

Risk

The term *risk* has a special meaning in statistics. It is defined as a situation where there are two or more possible outcomes and the probability associated with each outcome is known. In each of the two previous examples there is a risk situation. We could not know in advance what playing card would be dealt or what number would be rolled by the die. However, since the various probabilities could be computed, our definition of risk has been satisfied. Probability and risk are not restricted to gambling games. For example, in a particular engineering course, a student has computed the probability for each of the letter grades he might receive as follows:

Grade	Grade Point	Probability P(Grade)
A	4.0	0.10
B	3.0	0.30
C	2.0	0.25
D	1.0	0.20
F	0	0.15
		1.00

From the table we see that the grade with the highest probability is a B. This, therefore, is the most likely grade. We also see that there is a substantial probability that some grade other than a B will be received. And the probabilities indicate that if a B is not received, the grade will probably be something less than a B. But in saying that the most likely grade is a B, other outcomes are ignored. In the next section we will show that a composite statistic may be computed using all the data.

Expected Value

In the last example the most likely grade of B in an engineering class had a probability of 0.30. That is not a very high probability. In some other course, say a math class, we might estimate a probability of 0.65 of obtaining a B, again making the B the most likely grade. While a B is most likely in both classes, it is more certain in the math class.

We can compute a weighted mean to give a better understanding of the total situation as represented by various possible outcomes. When the probabilities are used as the weighting factors, the result is called the *expected value* and is written

$$\text{Expected value} = \text{Outcome}_A \times P(A) + \text{Outcome}_B \times P(B) + \cdots$$

Example **A.35**

An engineer wishes to determine the risk of fire loss for her $200,000 home. From a fire rating bureau she obtains the following data:

Outcome	Probability
No fire loss	0.986 in any year
$10,000 fire loss	0.010
40,000 fire loss	0.003
200,000 fire loss	0.001

Compute the expected fire loss in any year.

Solution

$$\text{Expected fire loss} = 10,000(0.010) + 40,000(0.003) + 200,000(0.001) = \$420$$

REFERENCE

Newnan, Donald G. *Engineering Economic Analysis*, 5th ed. Engineering Press, San Jose, CA, 1995.

INTEREST TABLES

Compound interest factors

$\frac{1}{2}\%$ $\frac{1}{2}\%$

	Single Payment		Uniform Payment Series				Uniform Gradient		
	Compound Amount Factor	Present Worth Factor	Sinking Fund Factor	Capital Recovery Factor	Compound Amount Factor	Present Worth Factor	Gradient Uniform Series	Gradient Present Worth	
	Find F Given P F/P	Find P Given F P/F	Find A Given F A/F	Find A Given P A/P	Find F Given A F/A	Find P Given A P/A	Find A Given G A/G	Find P Given G P/G	
n									n
1	1.005	.9950	1.0000	1.0050	1.000	0.995	0	0	1
2	1.010	.9901	.4988	.5038	2.005	1.985	0.499	0.991	2
3	1.015	.9851	.3317	.3367	3.015	2.970	0.996	2.959	3
4	1.020	.9802	.2481	.2531	4.030	3.951	1.494	5.903	4
5	1.025	.9754	.1980	.2030	5.050	4.926	1.990	9.803	5
6	1.030	.9705	.1646	.1696	6.076	5.896	2.486	14.660	6
7	1.036	.9657	.1407	.1457	7.106	6.862	2.980	20.448	7
8	1.041	.9609	.1228	.1278	8.141	7.823	3.474	27.178	8
9	1.046	.9561	.1089	.1139	9.182	8.779	3.967	34.825	9
10	1.051	.9513	.0978	.1028	10.228	9.730	4.459	43.389	10
11	1.056	.9466	.0887	.0937	11.279	10.677	4.950	52.855	11
12	1.062	.9419	.0811	.0861	12.336	11.619	5.441	63.218	12
13	1.067	.9372	.0746	.0796	13.397	12.556	5.931	74.465	13
14	1.072	.9326	.0691	.0741	14.464	13.489	6.419	86.590	14
15	1.078	.9279	.0644	.0694	15.537	14.417	6.907	99.574	15
16	1.083	.9233	.0602	.0652	16.614	15.340	7.394	113.427	16
17	1.088	.9187	.0565	.0615	17.697	16.259	7.880	128.125	17
18	1.094	.9141	.0532	.0582	18.786	17.173	8.366	143.668	18
19	1.099	.9096	.0503	.0553	19.880	18.082	8.850	160.037	19
20	1.105	9051	.0477	.0527	20.979	18.987	9.334	177.237	20
21	1.110	.9006	.0453	.0503	22.084	19.888	9.817	195.245	21
22	1.116	.8961	.0431	.0481	23.194	20.784	10.300	214.070	22
23	1.122	.8916	.0411	.0461	24.310	21.676	10.781	233.680	23
24	1.127	.8872	.0393	.0443	25.432	22.563	11.261	254.088	24
25	1.133	.8828	.0377	.0427	26.559	23.446	11.741	275.273	25
26	1.138	.8784	.0361	.0411	27.692	24.324	12.220	297.233	26
27	1.144	.8740	.0347	.0397	28.830	25.198	12.698	319.955	27
28	1.150	.8697	.0334	.0384	29.975	26.068	13.175	343.439	28
29	1.156	.8653	.0321	.0371	31.124	26.933	13.651	367.672	29
30	1.161	.8610	.0310	.0360	32.280	27.794	14.127	392.640	30
36	1.197	.8356	.0254	.0304	39.336	32.871	16.962	557.564	36
40	1.221	.8191	.0226	.0276	44.159	36.172	18.836	681.341	40
48	1.270	.7871	.0185	.0235	54.098	42.580	22.544	959.928	48
50	1.283	.7793	.0177	.0227	56.645	44.143	23.463	1 035.70	50
52	1.296	.7716	.0169	.0219	59.218	45.690	24.378	1 113.82	52
60	1.349	.7414	.0143	.0193	69.770	51.726	28.007	1 448.65	60
70	1.418	.7053	.0120	.0170	83.566	58.939	32.468	1 913.65	70
72	1.432	.6983	.0116	.0166	86.409	60.340	33.351	2 012.35	72
80	1.490	.6710	.0102	.0152	98.068	65.802	36.848	2 424.65	80
84	1.520	.6577	.00961	.0146	104.074	68.453	38.576	2 640.67	84
90	1.567	.6383	.00883	.0138	113.311	72.331	41.145	2 976.08	90
96	1.614	.6195	.00814	.0131	122.829	76.095	43.685	3 324.19	96
100	1.647	.6073	.00773	.0127	129.334	78.543	45.361	3 562.80	100
104	1.680	.5953	.00735	.0124	135.970	80.942	47.025	3 806.29	104
120	1.819	.5496	.00610	.0111	163.880	90.074	53.551	4 823.52	120
240	3.310	.3021	.00216	.00716	462.041	139.581	96.113	13 415.56	240
360	6.023	.1660	.00100	.00600	1 004.5	166.792	128.324	21 403.32	360
480	10.957	.0913	.00050	.00550	1 991.5	181.748	151.795	27 588.37	480

Compound interest factors

1%									1%
	Single Payment		**Uniform Payment Series**				**Uniform Gradient**		
	Compound Amount Factor	Present Worth Factor	Sinking Fund Factor	Capital Recovery Factor	Compound Amount Factor	Present Worth Factor	Gradient Uniform Series	Gradient Present Worth	
	Find F Given P F/P	Find P Given F P/F	Find A Given F A/F	Find A Given P A/P	Find F Given A F/A	Find P Given A P/A	Find A Given G A/G	Find P Given G P/G	
n									n
1	1.010	.9901	1.0000	1.0100	1.000	0.990	0	0	1
2	1.020	.9803	.4975	.5075	2.010	1.970	0.498	0.980	2
3	1.030	.9706	.3300	.3400	3.030	2.941	0.993	2.921	3
4	1.041	.9610	.2463	.2563	4.060	3.902	1.488	5.804	4
5	1.051	.9515	.1960	.2060	5.101	4.853	1.980	9.610	5
6	1.062	.9420	.1625	.1725	6.152	5.795	2.471	14.320	6
7	1.072	.9327	.1386	.1486	7.214	6.728	2.960	19.917	7
8	1.083	.9235	.1207	.1307	8.286	7.652	3.448	26.381	8
9	1.094	.9143	.1067	.1167	9.369	8.566	3.934	33.695	9
10	1.105	.9053	.0956	.1056	10.462	9.471	4.418	41.843	10
11	1.116	.8963	.0865	.0965	11.567	10.368	4.900	50.806	11
12	1.127	.8874	.0788	.0888	12.682	11.255	5.381	60.568	12
13	1.138	.8787	.0724	.0824	13.809	12.134	5.861	71.112	13
14	1.149	.8700	.0669	.0769	14.947	13.004	6.338	82.422	14
15	1.161	.8613	.0621	.0721	16.097	13.865	6.814	94.481	15
16	1.173	.8528	.0579	.0679	17.258	14.718	7.289	107.273	16
17	1.184	.8444	.0543	.0643	18.430	15.562	7.761	120.783	17
18	1.196	.8360	.0510	.0610	19.615	16.398	8.232	134.995	18
19	1.208	.8277	.0481	.0581	20.811	17.226	8.702	149.895	19
20	1.220	.8195	.0454	.0554	22.019	18.046	9.169	165.465	20
21	1.232	.8114	.0430	.0530	23.239	18.857	9.635	181.694	21
22	1.245	.8034	.0409	.0509	24.472	19.660	10.100	198.565	22
23	1.257	.7954	.0389	.0489	25.716	20.456	10.563	216.065	23
24	1.270	.7876	.0371	.0471	26.973	21.243	11.024	234.179	24
25	1.282	.7798	.0354	.0454	28.243	22.023	11.483	252.892	25
26	1.295	.7720	0339	.0439	29.526	22.795	11.941	272.195	26
27	1.308	.7644	.0324	.0424	30.821	23.560	12.397	292.069	27
28	1.321	.7568	.0311	.0411	32.129	24.316	12.852	312.504	28
29	1.335	.7493	.0299	.0399	33.450	25.066	13.304	333.486	29
30	1.348	.7419	.0287	.0387	34.785	25.808	13.756	355.001	30
36	1.431	.6989	.0232	.0332	43.077	30.107	16.428	494.620	36
40	1.489	.6717	.0205	.0305	48.886	32.835	18.178	596.854	40
48	1.612	.6203	.0163	.0263	61.223	37.974	21.598	820.144	48
50	1.645	.6080	.0155	.0255	64.463	39.196	22.436	879.417	50
52	1.678	.5961	.0148	.0248	67.769	40.394	23.269	939.916	52
60	1.817	.5504	.0122	.0222	81.670	44.955	26.533	1 192.80	60
70	2.007	.4983	.00993	.0199	100.676	50.168	30.470	1 528.64	70
72	2.047	.4885	.00955	.0196	104.710	51.150	31.239	1 597.86	72
80	2.217	.4511	.00822	.0182	121.671	54.888	34.249	1 879.87	80
84	2.307	.4335	.00765	.0177	130.672	56.648	35.717	2 023.31	84
90	2.449	.4084	.00690	.0169	144.863	59.161	37.872	2 240.56	90
96	2.599	.3847	.00625	.0163	159.927	61.528	39.973	2 459.42	96
100	2.705	.3697	.00587	.0159	170.481	63.029	41.343	2 605.77	100
104	2.815	.3553	.00551	.0155	181.464	64.471	42.688	2 752.17	104
120	3.300	.3030	.00435	.0143	230.039	69.701	47.835	3 334.11	120
240	10.893	.0918	.00101	.0110	989.254	90.819	75.739	6 878.59	240
360	35.950	.0278	.00029	.0103	3 495.0	97.218	89.699	8 720.43	360
480	118.648	.00843	.00008	.0101	11 764.8	99.157	95.920	9 511.15	480

Compound interest factors

	Single Payment		Uniform Payment Series				Uniform Gradient		
	Compound Amount Factor	Present Worth Factor	Sinking Fund Factor	Capital Recovery Factor	Compound Amount Factor	Present Worth Factor	Gradient Uniform Series	Gradient Present Worth	
	Find F Given P	Find P Given F	Find A Given F	Find A Given P	Find F Given A	Find P Given A	Find A Given G	Find P Given G	
n	F/P	P/F	A/F	A/P	F/A	P/A	A/G	P/G	n
1	1.015	.9852	1.0000	1.0150	1.000	0.985	0	0	1
2	1.030	.9707	.4963	.5113	2.015	1.956	0.496	0.970	2
3	1.046	.9563	.3284	.3434	3.045	2.912	0.990	2.883	3
4	1.061	.9422	.2444	.2594	4.091	3.854	1.481	5.709	4
5	1.077	.9283	.1941	.2091	5.152	4.783	1.970	9.422	5
6	1.093	.9145	.1605	.1755	6.230	5.697	2.456	13.994	6
7	1.110	.9010	.1366	.1516	7.323	6.598	2.940	19.400	7
8	1.126	.8877	.1186	.1336	8.433	7.486	3.422	25.614	8
9	1.143	.8746	.1046	.1196	9.559	8.360	3.901	32.610	9
10	1.161	.8617	.0934	.1084	10.703	9.222	4.377	40.365	10
11	1.178	.8489	.0843	.0993	11.863	10.071	4.851	48.855	11
12	1.196	.8364	.0767	.0917	13.041	10.907	5.322	58.054	12
13	1.214	.8240	.0702	.0852	14.237	11.731	5.791	67.943	13
14	1.232	.8118	.0647	.0797	15.450	12.543	6.258	78.496	14
15	1.250	.7999	.0599	.0749	16.682	13.343	6.722	89.694	15
16	1.269	.7880	.0558	.0708	17.932	14.131	7.184	101.514	16
17	1.288	.7764	.0521	.0671	19.201	14.908	7.643	113.937	17
18	1.307	.7649	.0488	.0638	20.489	15.673	8.100	126.940	18
19	1.327	.7536	.0459	.0609	21.797	16.426	8.554	140.505	19
20	1.347	.7425	.0432	.0582	23.124	17.169	9.005	154.611	20
21	1.367	.7315	.0409	.0559	24.470	17.900	9.455	169.241	21
22	1.388	.7207	.0387	.0537	25.837	18.621	9.902	184.375	22
23	1.408	.7100	.0367	.0517	27.225	19.331	10.346	199.996	23
24	1.430	.6995	.0349	.0499	28.633	20.030	10.788	216.085	24
25	1.451	.6892	.0333	.0483	30.063	20.720	11.227	232.626	25
26	1.473	.6790	.0317	.0467	31.514	21.399	11.664	249.601	26
27	1.495	.6690	.0303	.0453	32.987	22.068	12.099	266.995	27
28	1.517	.6591	.0290	.0440	34.481	22.727	12.531	284.790	28
29	1.540	.6494	.0278	.0428	35.999	23.376	12.961	302.972	29
30	1.563	.6398	.0266	.0416	37.539	24.016	13.388	321.525	30
36	1.709	.5851	.0212	.0362	47.276	27.661	15.901	439.823	36
40	1.814	.5513	.0184	.0334	54.268	29.916	17.528	524.349	40
48	2.043	.4894	.0144	.0294	69.565	34.042	20.666	703.537	48
50	2.105	.4750	.0136	.0286	73.682	35.000	21.428	749.955	50
52	2.169	.4611	.0128	.0278	77.925	35.929	22.179	796.868	52
60	2.443	.4093	.0104	.0254	96.214	39.380	25.093	988.157	60
70	2.835	.3527	.00817	.0232	122.363	43.155	28.529	1 231.15	70
72	2.921	.3423	.00781	.0228	128.076	43.845	29.189	1 279.78	72
80	3.291	.3039	.00655	.0215	152.710	46.407	31.742	1 473.06	80
84	3.493	.2863	.00602	.0210	166.172	47.579	32.967	1 568.50	84
90	3.819	.2619	.00532	.0203	187.929	49.210	34.740	1 709.53	90
96	4.176	.2395	.00472	.0197	211.719	50.702	36.438	1 847.46	96
100	4.432	.2256	.00437	.0194	228.802	51.625	37.529	1 937.43	100
104	4.704	.2126	.00405	.0190	246.932	52.494	38.589	2 025.69	104
120	5.969	.1675	.00302	.0180	331.286	55.498	42.518	2 359.69	120
240	35.632	.0281	.00043	.0154	2 308.8	64.796	59.737	3 870.68	240
360	212.700	.00470	.00007	.0151	14 113.3	66.353	64.966	4 310.71	360
480	1 269.7	.00079	.00001	.0150	84 577.8	66.614	66.288	4 415.74	480

$1\frac{1}{2}\%$ (top left and top right)

Compound interest factors

2%									**2%**
	Single Payment		Uniform Payment Series				Uniform Gradient		
	Compound Amount Factor	Present Worth Factor	Sinking Fund Factor	Capital Recovery Factor	Compound Amount Factor	Present Worth Factor	Gradient Uniform Series	Gradient Present Worth	
	Find F Given P	Find P Given F	Find A Given F	Find A Given P	Find F Given A	Find P Given A	Find A Given G	Find P Given G	
n	F/P	P/F	A/F	A/P	F/A	P/A	A/G	P/G	n
1	1.020	.9804	1.0000	1.0200	1.000	0.980	0	0	1
2	1.040	.9612	.4951	.5151	2.020	1.942	0.495	0.961	2
3	1.061	.9423	.3268	.3468	3.060	2.884	0.987	2.846	3
4	1.082	.9238	.2426	.2626	4.122	3.808	1.475	5.617	4
5	1.104	.9057	.1922	.2122	5.204	4.713	1.960	9.240	5
6	1.126	.8880	.1585	.1785	6.308	5.601	2.442	13.679	6
7	1.149	.8706	.1345	.1545	7.434	6.472	2.921	18.903	7
8	1.172	.8535	.1165	.1365	8.583	7.325	3.396	24.877	8
9	1.195	.8368	.1025	.1225	9.755	8.162	3.868	31.571	9
10	1.219	.8203	.0913	.1113	10.950	8.983	4.337	38.954	10
11	1.243	.8043	.0822	.1022	12.169	9.787	4.802	46.996	11
12	1.268	.7885	.0746	.0946	13.412	10.575	5.264	55.669	12
13	1.294	.7730	.0681	.0881	14.680	11.348	5.723	64.946	13
14	1.319	.7579	.0626	.0826	15.974	12.106	6.178	74.798	14
15	1.346	.7430	.0578	.0778	17.293	12.849	6.631	85.200	15
16	1.373	.7284	.0537	.0737	18.639	13.578	7.080	96.127	16
17	1.400	.7142	.0500	.0700	20.012	14.292	7.526	107.553	17
18	1.428	.7002	.0467	.0667	21.412	14.992	7.968	119.456	18
19	1.457	.6864	.0438	.0638	22.840	15.678	8.407	131.812	19
20	1.486	.6730	.0412	.0612	24.297	16.351	8.843	144.598	20
21	1.516	.6598	.0388	.0588	25.783	17.011	9.276	157.793	21
22	1.546	.6468	.0366	.0566	27.299	17.658	9.705	171.377	22
23	1.577	.6342	.0347	.0547	28.845	18.292	10.132	185.328	23
24	1.608	.6217	.0329	.0529	30.422	18.914	10.555	199.628	24
25	1.641	.6095	.0312	.0512	32.030	19.523	10.974	214.256	25
26	1.673	.5976	.0297	.0497	33.671	20.121	11.391	229.169	26
27	1.707	.5859	.0283	.0483	35.344	20.707	11.804	244.428	27
28	1.741	.5744	.0270	.0470	37.051	21.281	12.214	259.936	28
29	1.776	.5631	.0258	.0458	38.792	21.844	12.621	275.703	29
30	1.811	.5521	.0247	.0447	40.568	22.396	13.025	291.713	30
36	2.040	.4902	.0192	.0392	51.994	25.489	15.381	392.036	36
40	2.208	.4529	.0166	.0366	60.402	27.355	16.888	461.989	40
48	2.587	.3865	.0126	.0326	79.353	30.673	19.755	605.961	48
50	2.692	.3715	.0118	.0318	84.579	31.424	20.442	642.355	50
52	2.800	.3571	.0111	.0311	90.016	32.145	21.116	678.779	52
60	3.281	.3048	.00877	.0288	114.051	34.761	23.696	823.692	60
70	4.000	.2500	.00667	.0267	149.977	37.499	26.663	999.829	70
72	4.161	.2403	.00633	.0263	158.056	37.984	27.223	1 034.050	72
80	4.875	.2051	.00516	.0252	193.771	39.744	29.357	1 166.781	80
84	5.277	.1895	.00468	.0247	213.865	40.525	30.361	1 230.413	84
90	5.943	.1683	.00405	.0240	247.155	41.587	31.793	1 322.164	90
96	6.693	.1494	.00351	.0235	284.645	42.529	33.137	1 409.291	96
100	7.245	.1380	.00320	.0232	312.230	43.098	33.986	1 464.747	100
104	7.842	.1275	.00292	.0229	342.090	43.624	34.799	1 518.082	104
120	10.765	.0929	.00205	.0220	488.255	45.355	37.711	1 710.411	120
240	115.887	.00863	.00017	.0202	5 744.4	49.569	47.911	2 374.878	240
360	1 247.5	.00080	.00002	.0200	62 326.8	49.960	49.711	2 483.567	360
480	13 429.8	.00007		.0200	671 442.0	49.996	49.964	2 498.027	480

Compound interest factors

4%									4%
	Single Payment		**Uniform Payment Series**				**Uniform Gradient**		
	Compound Amount Factor	Present Worth Factor	Sinking Fund Factor	Capital Recovery Factor	Compound Amount Factor	Present Worth Factor	Gradient Uniform Series	Gradient Present Worth	
	Find F Given P	Find P Given F	Find A Given F	Find A Given P	Find F Given A	Find P Given A	Find A Given G	Find P Given G	
n	F/P	P/F	A/F	A/P	F/A	P/A	A/G	P/G	n
1	1.040	.9615	1.0000	1.0400	1.000	0.962	0	0	1
2	1.082	.9246	.4902	.5302	2.040	1.886	0.490	0.925	2
3	1.125	.8890	.3203	.3603	3.122	2.775	0.974	2.702	3
4	1.170	.8548	.2355	.2755	4.246	3.630	1.451	5.267	4
5	1.217	.8219	.1846	.2246	5.416	4.452	1.922	8.555	5
6	1.265	.7903	.1508	.1908	6.633	5.242	2.386	12.506	6
7	1.316	.7599	.1266	.1666	7.898	6.002	2.843	17.066	7
8	1.369	.7307	.1085	.1485	9.214	6.733	3.294	22.180	8
9	1.423	.7026	.0945	.1345	10.583	7.435	3.739	27.801	9
10	1.480	.6756	.0833	.1233	12.006	8.111	4.177	33.881	10
11	1.539	.6496	.0741	.1141	13.486	8.760	4.609	40.377	11
12	1.601	.6246	.0666	.1066	15.026	9.385	5.034	47.248	12
13	1.665	.6006	.0601	.1001	16.627	9.986	5.453	54.454	13
14	1.732	.5775	.0547	.0947	18.292	10.563	5.866	61.962	14
15	1.801	.5553	.0499	.0899	20.024	11.118	6.272	69.735	15
16	1.873	.5339	.0458	.0858	21.825	11.652	6.672	77.744	16
17	1.948	.5134	.0422	.0822	23.697	12.166	7.066	85.958	17
18	2.029	.4936	.0390	.0790	25.645	12.659	7.453	94.350	18
19	2.107	.4746	.0361	.0761	27.671	13.134	7.834	102.893	19
20	2.191	.4564	.0336	.0736	29.778	13.590	8.209	111.564	20
21	2.279	.4388	.0313	.0713	31.969	14.029	8.578	120.341	21
22	2.370	.4220	.0292	.0692	34.248	14.451	8.941	129.202	22
23	2.465	.4057	.0273	.0673	36.618	14.857	9.297	138.128	23
24	2.563	.3901	.0256	.0656	39.083	15.247	9.648	147.101	24
25	2.666	.3751	.0240	.0640	41.646	15.622	9.993	156.104	25
26	2.772	.3607	.0226	.0626	44.312	15.983	10.331	165.121	26
27	2.883	.3468	.0212	.0612	47.084	16.330	10.664	174.138	27
28	2.999	.3335	.0200	.0600	49.968	16.663	10.991	183.142	28
29	3.119	.3207	.0189	.0589	52.966	16.984	11.312	192.120	29
30	3.243	.3083	.0178	.0578	56.085	17.292	11.627	201.062	30
31	3.373	.2965	.0169	.0569	59.328	17.588	11.937	209.955	31
32	3.508	.2851	.0159	.0559	62.701	17.874	12.241	218.792	32
33	3.648	.2741	.0151	.0551	66.209	18.148	12.540	227.563	33
34	3.794	.2636	.0143	.0543	69.858	18.411	12.832	236.260	34
35	3.946	.2534	.0136	.0536	73.652	18.665	13.120	244.876	35
40	4.801	.2083	.0105	.0505	95.025	19.793	14.476	286.530	40
45	5.841	.1712	.00826	.0483	121.029	20.720	15.705	325.402	45
50	7.107	.1407	.00655	.0466	152.667	21.482	16.812	361.163	50
55	8.646	.1157	.00523	.0452	191.159	22.109	17.807	393.689	55
60	10.520	.0951	.00420	.0442	237.990	22.623	18.697	422.996	60
65	12.799	.0781	.00339	.0434	294.968	23.047	19.491	449.201	65
70	15.572	.0642	.00275	.0427	364.290	23.395	20.196	472.479	70
75	18.945	.0528	.00223	.0422	448.630	23.680	20.821	493.041	75
80	23.050	.0434	.00181	.0418	551.244	23.915	21.372	511.116	80
85	28.044	.0357	.00148	.0415	676.089	24.109	21.857	526.938	85
90	34.119	.0293	.00121	.0412	827.981	24.267	22.283	540.737	90
95	41.511	.0241	.00099	.0410	1 012.8	24.398	22.655	552.730	95
100	50.505	.0198	.00081	.0408	1 237.6	24.505	22.980	563.125	100

Compound interest factors

6%									6%
	Single Payment		**Uniform Payment Series**				**Uniform Gradient**		
	Compound Amount Factor	Present Worth Factor	Sinking Fund Factor	Capital Recovery Factor	Compound Amount Factor	Present Worth Factor	Gradient Uniform Series	Gradient Present Worth	
	Find F Given P F/P	Find P Given F P/F	Find A Given F A/F	Find A Given P A/P	Find F Given A F/A	Find P Given A P/A	Find A Given G A/G	Find P Given G P/G	
n									**n**
1	1.060	.943	1.0000	1.0600	1.000	0.943	0	0	1
2	1.124	.8900	.4854	.5454	2.060	1.833	0.485	0.890	2
3	1.191	.8396	.3141	.3741	3.184	2.673	0.961	2.569	3
4	1.262	.7921	.2286	.2886	4.375	3.465	1.427	4.945	4
5	1.338	.7473	.1774	.2374	5.637	4.212	1.884	7.934	5
6	1.419	.7050	.1434	.2034	6.975	4.917	2.330	11.459	6
7	1.504	.6651	.1191	.1791	8.394	5.582	2.768	15.450	7
8	1.594	.6274	.1010	.1610	9.897	6.210	3.195	19.841	8
9	1.689	.5919	.0870	.1470	11.491	6.802	3.613	24.577	9
10	1.791	.5584	.0759	.1359	13.181	7.360	4.022	29.602	10
11	1.898	.5268	.0668	.1268	14.972	7.887	4.421	34.870	11
12	2.012	.4970	.0593	.1193	16.870	8.384	4.811	40.337	12
13	2.133	.4688	.0530	.1130	18.882	8.853	5.192	45.963	13
14	2.261	.4423	.0476	.1076	21.015	9.295	5.564	51.713	14
15	2.397	.4173	.0430	.1030	23.276	9.712	5.926	57.554	15
16	2.540	.3936	.0390	.0990	25.672	10.106	6.279	63.459	16
17	2.693	.3714	.0354	.0954	28.213	10.477	6.624	69.401	17
18	2.854	.3503	.0324	.0924	30.906	10.828	6.960	75.357	18
19	3.026	.3305	.0296	.0896	33.760	11.158	7.287	81.306	19
20	3.207	.3118	.0272	.0872	36.786	11.470	7.605	87.230	20
21	3.400	.2942	.0250	.0850	39.993	11.764	7.915	93.113	21
22	3.604	.2775	.0230	.0830	43.392	12.042	8.217	98.941	22
23	3.820	.2618	.0213	.0813	46.996	12.303	8.510	104.700	23
24	4.049	.2470	.0197	.0797	50.815	12.550	8.795	110.381	24
25	4.292	.2330	.0182	.0782	54.864	12.783	9.072	115.973	25
26	4.549	.2198	.0169	.0769	59.156	13.003	9.341	121.468	26
27	4.822	.2074	.0157	.0757	63.706	13.211	9.603	126.860	27
28	5.112	.1956	.0146	.0746	68.528	13.406	9.857	132.142	28
29	5.418	.1846	.0136	.0736	73.640	13.591	10.103	137.309	29
30	5.743	.1741	.0126	.0726	79.058	13.765	10.342	142.359	30
31	6.088	.1643	.0118	.0718	84.801	13.929	10.574	147.286	31
32	6.453	.1550	.0110	.0710	90.890	14.084	10.799	152.090	32
33	6.841	.1462	.0103	.0703	97.343	14.230	11.017	156.768	33
34	7.251	.1379	.00960	.0696	104.184	14.368	11.228	161.319	34
35	7.686	.1301	.00897	.0690	111.435	11.498	11.432	165.743	35
40	10.286	.0972	.00646	.0665	154.762	15.046	12.359	185.957	40
45	13.765	.0727	.00470	.0647	212.743	15.456	13.141	203.109	45
50	18.420	.0543	.00344	.0634	290.335	15.762	13.796	217.457	50
55	24.650	.0406	.00254	.0625	394.171	15.991	14.341	229.322	55
60	32.988	.0303	.00188	.0619	533.126	16.161	14.791	239.043	60
65	44.145	.0227	.00139	.0614	719.080	16.289	15.160	246.945	65
70	59.076	.0169	.00103	.0610	967.928	16.385	15.461	253.327	70
75	79.057	.0126	.00077	.0608	1 300.9	16.456	15.706	258.453	75
80	105.796	.00945	.00057	.0606	1 746.6	16.509	15.903	262.549	80
85	141.578	.00706	.00043	.0604	2 343.0	16.549	16.062	265.810	85
90	189.464	.00528	.00032	.0603	3 141.1	16.579	16.189	268.395	90
95	253.545	.00394	.00024	.0602	4 209.1	16.601	16.290	270.437	95
100	339.300	.00295	.00018	.0602	5 638.3	16.618	16.371	272.047	100

Compound interest factors

8%									8%
	Single Payment		Uniform Payment Series				Uniform Gradient		
	Compound Amount Factor	Present Worth Factor	Sinking Fund Factor	Capital Recovery Factor	Compound Amount Factor	Present Worth Factor	Gradient Uniform Series	Gradient Present Worth	
	Find F Given P	Find P Given F	Find A Given F	Find A Given P	Find F Given A	Find P Given A	Find A Given G	Find P Given G	
n	F/P	P/F	A/F	A/P	F/A	P/A	A/G	P/G	n
1	1.080	.9259	1.0000	1.0800	1.000	0.926	0	0	1
2	1.166	.8573	.4808	.5608	2.080	1.783	0.481	0.857	2
3	1.260	.7938	.3080	.3880	3.246	2.577	0.949	2.445	3
4	1.360	.7350	.2219	.3019	4.506	3.312	1.404	4.650	4
5	1.469	.6806	.1705	.2505	5.867	3.993	1.846	7.372	5
6	1.587	.6302	.1363	.2163	7.336	4.623	2.276	10.523	6
7	1.714	.5835	.1121	.1921	8.923	5.206	2.694	14.024	7
8	1.851	.5403	.0940	.1740	10.637	5.747	3.099	17.806	8
9	1.999	.5002	.0801	.1601	12.488	6.247	3.491	21.808	9
10	2.159	.4632	.0690	.1490	14.487	6.710	3.871	25.977	10
11	2.332	.4289	.0601	.1401	16.645	7.139	4.240	30.266	11
12	2.518	.3971	.0527	.1327	18.977	7.536	4.596	34.634	12
13	2.720	.3677	.0465	.1265	21.495	7.904	4.940	39.046	13
14	2.937	.3405	.0413	.1213	24.215	8.244	5.273	43.472	14
15	3.172	.3152	.0368	.1168	27.152	8.559	5.594	47.886	15
16	3.426	.2919	.0330	.1130	30.324	8.851	5.905	52.264	16
17	3.700	.2703	.0296	.1096	33.750	9.122	6.204	56.588	17
18	3.996	.2502	.0267	.1067	37.450	9.372	6.492	60.843	18
19	4.316	.2317	.0241	.1041	41.446	9.604	6.770	65.013	19
20	4.661	.2145	.0219	.1019	45.762	9.818	7.037	69.090	20
21	5.034	.1987	.0198	.0998	50.423	10.017	7.294	73.063	21
22	5.437	.1839	.0180	.0980	55.457	10.201	7.541	76.926	22
23	5.871	.1703	.0164	.0964	60.893	10.371	7.779	80.673	24
24	6.341	.1577	.0150	.0950	66.765	10.529	8.007	84.300	24
25	6.848	.1460	.0137	.0937	73.106	10.675	8.225	87.804	25
26	7.396	.1352	.0125	.0925	79.954	10.810	8.435	91.184	26
27	7.988	.1252	.0114	.0914	87.351	10.935	8.636	94.439	27
28	8.627	.1159	.0105	.0905	95.339	11.051	8.829	97.569	28
29	9.317	.1073	.00962	.0896	103.966	11.158	9.013	100.574	29
30	10.063	.0994	.00883	.0888	113.283	11.258	9.190	103.456	30
31	10.868	.0920	.00811	.0881	123.346	11.350	9.358	106.216	31
32	11.737	.0852	.00745	.0875	134.214	11.435	9.520	108.858	32
33	12.676	.0789	.00685	.0869	145.951	11.514	9.674	111.382	33
34	13.690	.0730	.00630	.0863	158.627	11.587	9.821	113.792	34
35	14.785	.0676	.00580	.0858	172.317	11.655	9.961	116.092	35
40	21.725	.0460	.00386	.0839	259.057	11.925	10.570	126.042	40
45	31.920	.0313	.00259	.0826	386.506	12.108	11.045	133.733	45
50	46.902	.0213	.00174	.0817	573.771	12.233	11.411	139.593	50
55	68.914	.0145	.00118	.0812	848.925	12.319	11.690	144.006	55
60	101.257	.00988	.00080	.0808	1 253.2	12.377	11.902	147.300	60
65	148.780	.00672	.00054	.0805	1 847.3	12.416	12.060	149.739	65
70	218.607	.00457	.00037	.0804	2 720.1	12.443	12.178	151.533	70
75	321.205	.00311	.00025	.0802	4 002.6	12.461	12.266	152.845	75
80	471.956	.00212	.00017	.0802	5 887.0	12.474	12.330	153.800	80
85	693.458	.00144	.00012	.0801	8 655.7	12.482	12.377	154.492	85
90	1 018.9	.00098	.00008	.0801	12 724.0	12.488	12.412	154.993	90
95	1 497.1	.00067	.00005	.0801	18 701.6	12.492	12.437	155.352	95
100	2 199.8	.00045	.00004	.0800	27 484.6	12.494	12.455	155.611	100

Compound interest factors

	Single Payment		Uniform Payment Series				Uniform Gradient		
	Compound Amount Factor	Present Worth Factor	Sinking Fund Factor	Capital Recovery Factor	Compound Amount Factor	Present Worth Factor	Gradient Uniform Series	Gradient Present Worth	
n	Find F Given P F/P	Find P Given F P/F	Find A Given F A/F	Find A Given P A/P	Find F Given A F/A	Find P Given A P/A	Find A Given G A/G	Find P Given G P/G	n
1	1.100	.9091	1.0000	1.1000	1.000	0.909	0	0	1
2	1.210	.8264	.4762	.5762	2.100	1.736	0.476	0.826	2
3	1.331	.7513	.3021	.4021	3.310	2.487	0.937	2.329	3
4	1.464	.6830	.2155	.3155	4.641	3.170	1.381	4.378	4
5	1.611	.6209	.1638	.2638	6.105	3.791	1.810	6.862	5
6	1.772	.5645	.1296	.2296	7.716	4.355	2.224	9.684	6
7	1.949	.5132	.1054	.2054	9.487	4.868	2.622	12.763	7
8	2.144	.4665	.0874	.1874	11.436	5.335	3.004	16.029	8
9	2.358	.4241	.0736	.1736	13.579	5.759	3.372	19.421	9
10	2.594	.3855	.0627	.1627	15.937	6.145	3.725	22.891	10
11	2.853	.3505	.0540	.1540	18.531	6.495	4.064	26.396	11
12	3.138	.3186	.0468	.1468	21.384	6.814	4.388	29.901	12
13	3.452	.2897	.0408	.1408	24.523	7.103	4.699	33.377	13
14	3.797	.2633	.0357	.1357	27.975	7.367	4.996	36.801	14
15	4.177	.2394	.0315	.1315	31.772	7.606	5.279	40.152	15
16	4.595	.2176	.0278	.1278	35.950	7.824	5.549	43.416	16
17	5.054	.1978	.0247	.1247	40.545	8.022	5.807	46.582	17
18	5.560	.1799	.0219	.1219	45.599	8.201	6.053	49.640	18
19	6.116	.1635	.0195	.1195	51.159	8.365	6.286	52.583	19
20	6.728	.1486	.0175	.1175	57.275	8.514	6.508	55.407	20
21	7.400	.1351	.0156	.1156	64.003	8.649	6.719	58.110	21
22	8.140	.1228	.0140	.1140	71.403	8.772	6.919	60.689	22
23	8.954	.1117	.0126	.1126	79.543	8.883	7.108	63.146	24
24	9.850	.1015	.0113	.1113	88.497	8.985	7.288	65.481	24
25	10.835	.0923	.0102	.1102	98.347	9.077	7.458	67.696	25
26	11.918	.0839	.00916	.1092	109.182	9.161	7.619	69.794	26
27	13.110	.0763	.00826	.1083	121.100	9.237	7.770	71.777	27
28	14.421	.0693	.00745	.1075	134.210	9.307	7.914	73.650	28
29	15.863	.0630	.00673	.1067	148.631	9.370	8.049	75.415	29
30	17.449	.0573	.00608	.1061	164.494	9.427	8.176	77.077	30
31	19.194	.0521	.00550	.1055	181.944	9.479	8.296	78.640	31
32	21.114	.0474	.00497	.1050	201.138	9.526	8.409	80.108	32
33	23.225	.0431	.00450	.1045	222.252	9.569	8.515	81.486	33
34	25.548	.0391	.00407	.1041	245.477	9.609	8.615	82.777	34
35	28.102	.0356	.00369	.1037	271.025	9.644	8.709	83.987	35
40	45.259	.0221	.00226	.1023	442.593	9.779	9.096	88.953	40
45	72.891	.0137	.00139	.1014	718.905	9.863	9.374	92.454	45
50	117.391	.00852	.00086	.1009	1 163.9	9.915	9.570	94.889	50
55	189.059	.00529	.00053	.1005	1 880.6	9.947	9.708	96.562	55
60	304.482	.00328	.00033	.1003	3 034.8	9.967	9.802	97.701	60
65	490.371	.00204	.00020	.1002	4 893.7	9.980	9.867	98.471	65
70	789.748	.00127	.00013	.1001	7 887.5	9.987	9.911	98.987	70
75	1 271.9	.00079	.00008	.1001	12 709.0	9.992	9.941	99.332	75
80	2 048.4	.00049	.00005	.1000	20 474.0	9.995	9.961	99.561	80
85	3 229.0	.00030	.00003	.1000	32 979.7	9.997	9.974	99.712	85
90	5 313.0	.00019	.00002	.1000	53 120.3	9.998	9.983	99.812	90
95	8 556.7	.00012	.00001	.1000	85 556.9	9.999	9.989	99.877	95
100	13 780.6	.00007	.00001	.1000	137 796.3	9.999	9.993	99.920	100

Compound interest factors

	Single Payment		Uniform Payment Series				Uniform Gradient		
	Compound Amount Factor	Present Worth Factor	Sinking Fund Factor	Capital Recovery Factor	Compound Amount Factor	Present Worth Factor	Gradient Uniform Series	Gradient Present Worth	
	Find F Given P F/P	Find P Given F P/F	Find A Given F A/F	Find A Given P A/P	Find F Given A F/A	Find P Given A P/A	Find A Given G A/G	Find P Given G P/G	
n									n
1	1.120	.8929	1.0000	1.1200	1.000	0.893	0	0	1
2	1.254	.7972	.4717	.5917	2.120	1.690	0.472	0.797	2
3	1.405	.7118	.2963	.4163	3.374	2.402	0.925	2.221	3
4	1.574	.6355	.2092	.3292	4.779	3.037	1.359	4.127	4
5	1.762	.5674	.1574	.2774	6.353	3.605	1.775	6.397	5
6	1.974	.5066	.1232	.2432	8.115	4.111	2.172	8.930	6
7	2.211	.4523	.0991	.2191	10.089	4.564	2.551	11.644	7
8	2.476	.4039	.0813	.2013	12.300	4.968	2.913	14.471	8
9	2.773	.3606	.0677	.1877	14.776	5.328	3.257	17.356	9
10	3.106	.3220	.0570	.1770	17.549	5.650	3.585	20.254	10
11	3.479	.2875	.0484	.1684	20.655	5.938	3.895	23.129	11
12	3.896	.2567	.0414	.1614	24.133	6.194	4.190	25.952	12
13	4.363	.2292	.0357	.1557	28.029	6.424	4.468	28.702	13
14	4.887	.2046	.0309	.1509	32.393	6.628	4.732	31.362	14
15	5.474	.1827	.0268	.1468	37.280	6.811	4.980	33.920	15
16	6.130	.1631	.0234	.1434	42.753	6.974	5.215	36.367	16
17	6.866	.1456	.0205	.1405	48.884	7.120	5.435	38.697	17
18	7.690	.1300	.0179	.1379	55.750	7.250	5.643	40.908	18
19	8.613	.1161	.0158	.1358	63.440	7.366	5.838	42.998	19
20	9.646	.1037	.0139	.1339	72.052	7.469	6.020	44.968	20
21	10.804	.0926	.0122	.1322	81.699	7.562	6.191	46.819	21
22	12.100	.0826	.0108	.1308	92.503	7.645	6.351	48.554	22
23	13.552	.0738	.00956	.1296	104.603	7.718	6.501	50.178	24
24	15.179	.0659	.00846	.1285	118.155	7.784	6.641	51.693	24
25	17.000	.0588	.00750	.1275	133.334	7.843	6.771	53.105	25
26	19.040	.0525	.00665	.1267	150.334	7.896	6.892	54.418	26
27	21.325	.0469	.00590	.1259	169.374	7.943	7.005	55.637	27
28	23.884	.0419	.00524	.1252	190.699	7.984	7.110	56.767	28
29	26.750	.0374	.00466	.1247	214.583	8.022	7.207	57.814	29
30	29.960	.0334	.00414	.1241	241.333	8.055	7.297	58.782	30
31	33.555	.0298	.00369	.1237	271.293	8.085	7.381	59.676	31
32	37.582	.0266	.00328	.1233	304.848	8.112	7.459	60.501	32
33	42.092	.0238	.00292	.1229	342.429	8.135	7.530	61.261	33
34	47.143	.0212	.00260	.1226	384.521	8.157	7.596	61.961	34
35	52.800	.0189	.00232	.1223	431.663	8.176	7.658	62.605	35
40	93.051	.0107	.00130	.1213	767.091	8.244	7.899	65.116	40
45	163.988	.00610	.00074	.1207	1 358.2	8.283	8.057	66.734	45
50	289.002	.00346	.00042	.1204	2 400.0	8.304	8.160	67.762	50
55	509.321	.00196	.00024	.1202	4 236.0	8.317	8.225	68.408	55
60	897.597	.00111	.00013	.1201	7 471.6	8.324	8.266	68.810	60
65	1 581.9	.00063	.00008	.1201	13 173.9	8.328	8.292	69.058	65
70	2 787.8	.00036	.00004	.1200	23 223.3	8.330	8.308	69.210	70
75	4 913.1	.00020	.00002	.1200	40 933.8	8.332	8.318	69.303	75
80	8 658.5	.00012	.00001	.1200	72 145.7	8.332	8.324	69.359	80
85	15 259.2	.00007	.00001	.1200	127 151.7	8.333	8.328	69.393	85
90	26 891.9	.00004		.1200	224 091.1	8.333	8.330	69.414	90
95	47 392.8	.00002		.1200	394 931.4	8.333	8.331	69.426	95
100	83 522.3	.00001		.1200	696 010.5	8.333	8.332	69.434	100

12% ... 12%

Compound interest factors

18%										18%
	Single Payment		**Uniform Payment Series**				**Uniform Gradient**			
	Compound Amount Factor	Present Worth Factor	Sinking Fund Factor	Capital Recovery Factor	Compound Amount Factor	Present Worth Factor	Gradient Uniform Series	Gradient Present Worth		
	Find F Given P	Find P Given F	Find A Given F	Find A Given P	Find F Given A	Find P Given A	Find A Given G	Find P Given G		
n	F/P	P/F	A/F	A/P	F/A	P/A	A/G	P/G	n
1	1.180	.8475	1.0000	1.1800	1.000	0.847	0	0	1
2	1.392	.7182	.4587	.6387	2.180	1.566	0.459	0.718	2
3	1.643	.6086	.2799	.4599	3.572	2.174	0.890	1.935	3
4	1.939	.5158	.1917	.3717	5.215	2.690	1.295	3.483	4
5	2.288	.4371	.1398	.3198	7.154	3.127	1.673	5.231	5
6	2.700	.3704	.1059	.2859	9.442	3.498	2.025	7.083	6
7	3.185	.3139	.0824	.2624	12.142	3.812	2.353	8.967	7
8	3.759	.2660	.0652	.2452	15.327	4.078	2.656	10.829	8
9	4.435	.2255	.0524	.2324	19.086	4.303	2.936	12.633	9
10	5.234	.1911	.0425	.2225	23.521	4.494	3.194	14.352	10
11	6.176	.1619	.0348	.2148	28.755	4.656	3.430	15.972	11
12	7.288	.1372	.0286	.2086	34.931	4.793	3.647	17.481	12
13	8.599	.1163	.0237	.2037	42.219	4.910	3.845	18.877	13
14	10.147	.0985	.0197	.1997	50.818	5.008	4.025	20.158	14
15	11.974	.0835	.0164	.1964	60.965	5.092	4.189	21.327	15
16	14.129	.0708	.0137	.1937	72.939	5.162	4.337	22.389	16
17	16.672	.0600	.0115	.1915	87.068	5.222	4.471	23.348	17
18	19.673	.0508	.00964	.1896	103.740	5.273	4.592	24.212	18
19	23.214	.0431	.00810	.1881	123.413	5.316	4.700	24.988	19
20	27.393	.0365	.00682	.1868	146.628	5.353	4.798	25.681	20
21	32.324	.0309	.00575	.1857	174.021	5.384	4.885	26.330	21
22	38.142	.0262	.00485	.1848	206.345	5.410	4.963	26.851	22
23	45.008	.0222	.00409	.1841	244.487	5.432	5.033	27.339	24
24	53.109	.0188	.00345	.1835	289.494	5.451	5.095	27.772	24
25	62.669	.0160	.00292	.1829	342.603	5.467	5.150	28.155	25
26	73.949	.0135	.00247	.1825	405.272	5.480	5.199	28.494	26
27	87.260	.0115	.00209	.1821	479.221	5.492	5.243	28.791	27
28	102.966	.00971	.00177	.1818	566.480	5.502	5.281	29.054	28
29	121.500	.00823	.00149	.1815	669.447	5.510	5.315	29.284	29
30	143.370	.00697	.00126	.1813	790.947	5.517	5.345	29.486	30
31	169.177	.00591	.00107	.1811	934.317	5.523	5.371	29.664	31
32	199.629	.00501	.00091	.1809	1 103.5	5.528	5.394	29.819	32
33	235.562	.00425	.00077	.1808	1 303.1	5.532	5.415	29.955	33
34	277.963	.00360	.00065	.1806	1 538.7	5.536	5.433	30.074	34
35	327.997	.00305	.00055	.1806	1 816.6	5.539	5.449	30.177	35
40	750.377	.00133	.00024	.1802	4 163.2	5.548	5.502	30.527	40
45	1 716.7	.00058	.00010	.1801	9 531.6	5.552	5.529	30.701	45
50	3 927.3	.00025	.00005	.1800	21 813.0	5.554	5.543	30.786	50
55	8 984.8	.00011	.00002	.1800	49 910.1	5.555	5.549	30.827	55
60	20 555.1	.00005	.00001	.1800	114 189.4	5.555	5.553	30.846	60
65	47 025.1	.00002		.1800	261 244.7	5.555	5.554	30.856	65
70	107 581.9	.00001		.1800	597 671.7	5.556	5.555	30.860	70
75	46 122.1				1 367 339.2	5.556	5.555	30.862	75
100	15 424 131.9				85 689 616.2	5.556	5.555	30.864	100